Nora Haydee Quispe Bellido
Yesica Magnolia Mamani Arpasi
Juan de Dios H. Ticona Quispe

Occupational safety and health management system

Nora Haydee Quispe Bellido
Yesica Magnolia Mamani Arpasi
Juan de Dios H. Ticona Quispe

Occupational safety and health management system

Planning proposal to manage occupational risk control in a Peruvian company

ScienciaScripts

Imprint

Any brand names and product names mentioned in this book are subject to trademark, brand or patent protection and are trademarks or registered trademarks of their respective holders. The use of brand names, product names, common names, trade names, product descriptions etc. even without a particular marking in this work is in no way to be construed to mean that such names may be regarded as unrestricted in respect of trademark and brand protection legislation and could thus be used by anyone.

Cover image: www.ingimage.com

This book is a translation from the original published under ISBN 978-613-9-02464-3.

Publisher:
Sciencia Scripts
is a trademark of
Dodo Books Indian Ocean Ltd. and OmniScriptum S.R.L publishing group

120 High Road, East Finchley, London, N2 9ED, United Kingdom
Str. Armeneasca 28/1, office 1, Chisinau MD-2012, Republic of Moldova, Europe
Printed at: see last page
ISBN: 978-620-7-62037-1

OCCUPATIONAL HEALTH AND SAFETY MANAGEMENT SYSTEM

Planning proposal to manage occupational risk control in a Peruvian company

Nora Haydeé Quispe Bellido

Yesica Magnolia Mamani Arpasi

Juan de Dios Hermogenes Ticona Quispe

Table of contents

Introduction

In recent years, the Occupational Health and Safety Management System (OHSMS) has become one of the most widely used systems among production and service organizations to enhance healthy work environments by providing an environment that can effectively identify and control health risks, reduce accidents, support the enforcement of current legal health requirements and improve performance[1] .

It is important to note that the OHSAS 18001 standard was developed to prevent occupational hazards[2] . The specification of OSHMS standards enables companies to control occupational health and safety risks with a view to continuous improvement[3] .

In this sense, a proposal is presented for the design of a management system in Peruvian companies, so that different organizations can use it and contribute to the continuous improvement of their performance.

Therefore, this book is structured in five chapters. The first chapter contains information on occupational health and safety, as well as measures and working conditions to prevent risks. The second chapter provides a review of the legal framework related to risk management, particularly in relation to the OHSAS 18001

---

[1] Ministry of Labor, Employment and Social Security. Occupational Health and Safety (OSH). *Aportes para una cultura de la prevención.* Argentina, ILO, 2014, available at [https://www.ilo.org/wcmsp5/groups/public/@americas/@@ro-lima/@ilo-buenos_aires/documents/publication/wcms_248685.pdf].

[2] International Labor Organization. *Occupational safety and health in Peru. Una mirada desde los convenios internacionales del trabajo no ratificado.* Peru, ILO, 2022, available at [https://www.ilo.org/wcmsp5/groups/public/---americas/---ro-lima/documents/publication/wcms_884854.pdf].

[3] Spanish Association for Standardization and Certification (Ed.). *OHSAS 18001:2007. Occupational health and safety management systems - Requirements.* Madrid, AENOR, 2007, disponible en [https://infomadera.net/uploads/descargas/archivo_49_Sistemas%20de%20gesti%C3%B3n%20de%2 0seguridad%20y%20salud%20OHSAS%2018001-2007.pdf].

standard. The third chapter describes studies focused on risk management and situations that have arisen in the country.

This is followed by a chapter on the design of an occupational health and safety system for a sanitation company based on a diagnosis that reveals its situation and what needs to be improved for the well-being of its personnel. Finally, the fifth chapter deals with the importance of developing and implementing an emergency plan in companies worldwide, to prevent risks caused by nature or by man, in order to reduce occupational accidents and increase the country's productivity.

CHAPTER ONE :

OCCUPATIONAL HEALTH AND SAFETY: RISK CONTROL MEASURES

I. Historical evolution of occupational health and safety

As social, technological, legal and ethical changes have occurred in today's environment, security has also evolved.

Occupational health consists of three main areas: occupational medicine, occupational health and industrial safety. "Through occupational health it is intended to improve and maintain the quality of life and health of workers and serve as an instrument to improve the quality, productivity and efficiency of companies"[4] .

Since 1997, a period of innovation and paradigm change has begun in the mining industry, since it has been possible to achieve a controlled reduction in the number of deaths due to accidents[5] . In this sense, several theories have been generated, among them:

— Occupational health and safety management became the responsibility of the companies and then came under government control as regulated by legal provisions.

— Safety consists of exercising control over risks, not in committing them (accidents).

---

[4] Fernando Henao Robledo. *Salud Ocupacional: conceptos básicos*, 2nd ed., Bogotá, Colombia, Ecoe Ediciones, 2010, p. 33.

[5] Arturo Miguel Pérez Cabrera. "Incidence of accident risks on the operating costs of non-metallic resource mining concessions in Patapo - Lambayeque" (undergraduate thesis). Pimentel, Peru, Universidad Señor de Sipán, 2019, available at [https://repositorio.uss.edu.pe/handle/20.500.12802/5551].

— Both business organizations and their employees have a responsibility to assess and identify risks at work in order to implement actions to address them.

— Work is done in teams.

— Corrective actions are used in order to anticipate risk situations.

— The person in charge of occupational safety is the owner, not the safety engineer (organizer, consultant and safety manager).

II. Industrial safety

Mancera *et al.*[6] define industrial safety as any activity aimed at preventing, identifying and controlling the causes of industrial accidents. Its purpose is to anticipate particular and general risk factors present in the workplace that are actual or potential causes of accidents.

Therefore, it is a multidisciplinary field in charge of reducing the risk of accidents within a company, whose activities, whatever their nature, present intrinsic hazards that require adequate management[7].

According to Cortés[8], the fundamental risks in this sector are related to accidents and illnesses that have a significant impact inside and outside the company where the accident occurred. Therefore, industrial safety must not only comply with the legal regulations in force regarding occupational health and safety, but also carry out the respective requirements set by the companies according to their activities;

---

[6] Mario Mancera Fernández, María Teresa Mancera Ruíz, Mario Ramón Mancera Ruíz and Juan Ricardo Mancera Ruíz. *Safety and industrial hygiene. Gestión de riesgos.* Colombia, Editorial Alfaomega, 2012, available at [https://ashconsultores.com.ar/wp-content/uploads/2019/06/Libro_Seguridad_e_Higiene_industrial_ges.pdf], p. 12.

[7] Juan Daniel Hernández Domínguez. "Occupational safety to prevent accidents in industry." *Ibero-American Journal of Academic Production and Educational Management*, vol. 5, no. 10, 2018, pp. 1-9, available at [https://www.pag.org.mx/index.php/PAG/article/view/773/1109].

[8] José María Cortés Díaz. Seguridad e higiene del trabajo: Técnicas de prevención de riesgos laborales, 12th ed., Madrid, Spain, Tébar, 2012.

employees must have adequate clothing and other essential elements, they also require medical supervision, technical controls and training on risk management.

It should be noted that safety in the industrial sector is relative, since it does not guarantee that no accident will occur[9] .

III. Occupational health and safety

Health in the work environment refers to diseases or hazardous situations linked or not to their work in a specific company, as well as those linked to an environment external to the work environment[10] .

According to Pastor *et al.*[11] , the implementation of a risk assessment should in no way be considered as a bureaucratic reinforcement, which is not an end in itself, but as a means to achieve corporate goals, identify risks and, if essential, employ preventive measures. It is therefore essential that:

— Risks are eliminated or minimized by means of preventive measures at source, organization, that protect individually or collectively, or even during training and the provision of data to personnel.

— Work spaces, techniques, organizational structure and workers' health are repeatedly monitored and protected[12] .

This is one of the key aspects in work activities and are understood as interdependent elements that establish a safety and health policy in the work

---

[9] Jaime Antonio Ortega Alarcón, Jorge Rafael Rodríguez López and Hugo Hernández Palma. "Importance of worker safety in the compliance of processes, procedures and functions." *Academia & Derecho Journal*, vol. 8, no. 14, 2017, pp. 155-176, available at [https://dialnet.unirioja.es/servlet/articulo?codigo=6713605].

[10] Eduardo Raffo Lecca. *Introducción a la seguridad y salud en el trabajo*. Lima, Colecciones Jovic, 2016.

[11] Andrés Pastor Fernández, Manuel Otero Mateo, José María Portela Núñez and José Luis Viguera Cebrián. *Manual de prácticas de seguridad en el trabajo*. Spain, Editorial UCA, 2016.

[12] Rafael Rodríguez Mesa. *Sistema general de riesgos laborales*, 3rd ed., Colombia, Universidad del Norte, 2017.

environment, promoting a culture of risk prevention and, in turn, optimizing work spaces during the execution of the task[13] .

A. Work and health

Man has used the existing goods for his own benefit since his appearance, in order to satisfy his nutritional and protection needs.

Thus, during the evolution of mankind, societies were formed, therefore, the natural goods used not only satisfied individual needs, but other uses were created, for example, for recreational activities. Then these generated needs and population growth, as well as nature's own limitations, require optimizing the use of these resources[14] .

Thus, the use of natural resources is redesigned to achieve a higher yield. This conversion process is called work.

However, this conversion may exceed the capabilities of a person and, because of its uncontrolled nature, pose a risk to the health of the person. Such a source of health risk is called a hazard.

ENEL[15] indicates that hazard is the quality of a situation, material or equipment that can harm individuals, the environment or cultural heritage. While occupational risk is defined as the probability that an employee suffers injuries due to the work performed. The risk is categorized according to its severity, it is evaluated together with the occurrence and its severity.

---

[13] ENEL. *Internal regulations for occupational health and safety*. Lima, Peru, ENEL, 2021, available at [https://www.enel.pe/content/dam/enel-pe/sostenibilidad/sistemas-de-gesti%C3%B3n/enel-distribuci%C3%B3n/sistemas-de-gesti%C3%B3n-actualizados/Reglamento%20Interno%20de%20Seguridad%20y%20Salud%20en%20el%20Trabajo%20-%20V9.pdf].

[14] Maritza Edilma Sabogal Barbosa and Félix Eduardo Rodríguez Medina. "Competencies of the technician in the work area. Un análisis del programa técnico manejo de prevención de riesgos laborales de la Fundación Universitaria San Mateo." *Plataforma Abierta de Libros y Memorias Académicas*, vol. 1, 2019, pp. 9-36, available at [https://cipres.sanmateo.edu.co/ojs/index.php/libros/article/view/396].

[15] ENEL. *Reglamento interno de seguridad y salud en el trabajo*, cit.

IV. National occupational health and safety system

It consists of all those actors and rules regulated within a country and within national legal frameworks in order to facilitate how to prevent risks in the work environment and to promote working conditions, such as the development of legal standards, inspections, training, promotion and support and information, registration, health care and insurance, participation and consultation with employees to identify and repeatedly review activities to ensure employee protection; Meanwhile, for employers, priority is given to the execution of each process and competitive strategies are promoted to ensure a position in the market[16] .

V. Occupational health and safety supervisor

Refers to a worker who has the necessary skills and competencies to perform supervisory duties, chosen by the managers of an organization or other entity, to a maximum of twenty employees[17] . It is in charge of verifying the indices related to accidents in the workplace. Some of the rates evaluated are[18] :

— *Frequency index (FI)*. This refers to the number of fatal and disabling accidents per million hours worked. The calculation is made with this formula:

$$IF = \frac{N°\ Accidentes\ \times 1\ 000\ 000\ (N°\ Accidentes = Incap. + fatal)}{Horas\ hombre\ trabajadas}$$

---

[16] Ministry of Labor and Employment Promotion. *Ley de Seguridad y Salud en el Trabajo, its regulations and amendments*. Lima, Peru, MTPE, 2017, available at [https://cdn.www.gob.pe/uploads/document/file/349382/LEY_DE_SEGURIDAD_Y_SALUD_EN_E L_TRABAJO.pdf].

[17] Idem.

[18] Liliana Rosalinda Agustini Paredes, Pedro Pablo Rosales López and Anwar Julio Yaroin Achachagua. *Ratios of accidentability*. Lima, Perú, Universidad Nacional Mayor de San Marcos, 2021, available at [https://industrial.unmsm.edu.pe/wp-content/uploads/2021/04/PSEG103-Ratios-de-Accidentabilidad.pdf].

— *Severity Index (SI)*. It consists of the number of days lost or billed per million hours worked. The calculation is made using this formula:

$$IS = \frac{N° \, D\acute{\imath}as \, perdidos \, o \, cargados \, \times 1\,000\,000}{Horas \, hombre \, trabajadas}$$

— *Accidentability Index (AI)*. It is a measure that combines the IF and the IS, therefore, the resulting product of the aforementioned indexes is divided by one thousand.

$$IA = \frac{IF \, \times IS}{1000}$$

VI. Working Conditions

It is that characteristic (local, facilities, equipment, etc.) that significantly affects the occurrence of risks inside or outside the work environment of a worker[19].

The pollutants and natural elements found in the occupational environment and the procedures for their use also influence the generation of risks.

Likewise, all other characteristics of the work, including those related to its organization and organization, affect the magnitude of the risks to which the worker is exposed.

The characteristics that working conditions must have are premises, installations, equipment, products, tools, agents (physical, chemical, biological and ergonomic, processes and organization).

Similarly, it is important to integrate:

— the work center, especially its location and accessibility;

— the overall activity of the company;

— the activity of neighboring companies;

---

[19] Mancera Fernández, Mancera Ruíz, Mancera Ruíz and Mancera Ruíz. *Seguridad e higiene industrial. Gestión de riesgos*, cit.

— simultaneous non-standard activities (project implementation, extensions and changes in general);

— and emergency situations.

Regarding emergency situations:

— fires;

— explosions;

— leakage of noxious gases;

— uncontrolled spills of hazardous products

— and ionizing radiation.

A. Damages arising from work

According to WHO, "health is a state of complete physical, mental and social well-being and not merely the absence of disease or infirmity"[20] .

The most important aspect of this definition is the triple dimension it proposes, and the importance of all three being in balance. Likewise, this concept represents the fact as a social element that contributes to the progress of a society and the development of the people who make it up.

Figure 1. Schematic representation of the concept of health.

---

[20] Gustavo Alcántara Moreno. "The World Health Organization's definition of health and interdisciplinarity." *Sapiens, Revista Universitaria de Investigación*, vol. 9, no. 1, 2008, pp. 93-107, available at [https://www.redalyc.org/pdf/410/41011135004.pdf], p. 96.

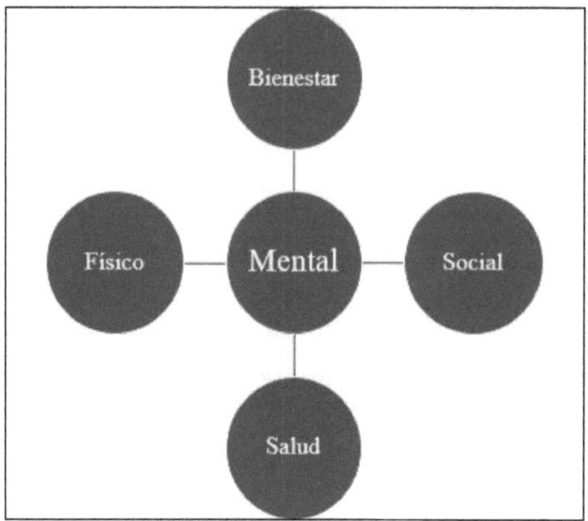

Source: Gustavo Alcántara Moreno. "La definición de salud de la Organización Mundial de la Salud y la interdisciplinariedad", cit.

The materialization of a risk can cause damage to health, the most noticeable manifestations of which are accidents or diseases. In addition, the fundamental characteristics of these manifestations of damage are:

— *In the accident*, the damage to health occurs suddenly and unexpectedly. It is an immediate and most evident indicator of poor working conditions.

— *In the disease*, the damage is a gradual and slow deterioration of the worker's health caused by chronic exposure to adverse conditions during the performance of the work.

B. Accident rate pyramid

The accident rate pyramid has been studied by George Germain and Frank Bird, who point out that, for every 600 incidents, there are 30 minor accidents, 10 serious accidents and one serious accident[21] .

---

[21] Raffo Lecca. *Introducción a la seguridad y salud en el trabajo*, cit.

In this pyramid, certain terms are used in reference to incidents, the definitions of which are given below:

— *Incident.* Refers to an unintentional event linked to work, which may or may not involve a health problem. This term, *lato sensu*, includes the varieties of industrial accidents[22] .
— *Causes of incidents.* These are one or more related events that occur to generate an accident. They are divided into:
  • *Lack of control.* This refers to failures, absences or weaknesses in the occupational health and safety management system.
  • *Basic causes.* Referred to personal factors and work factors.
  • *Immediate causes.* Refers to substandard acts or conditions.

Figure 2. The accident rate pyramid

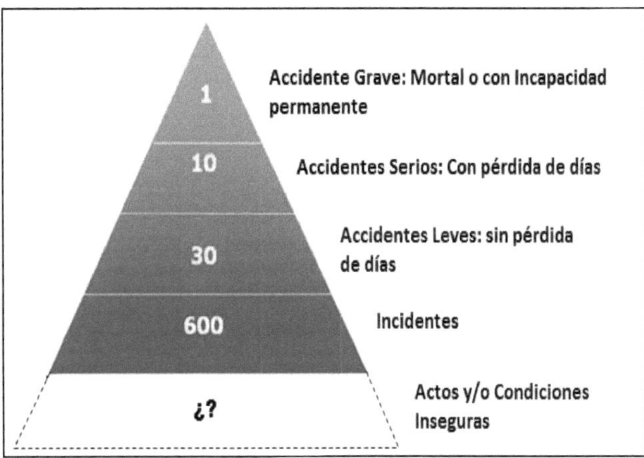

---

[22] Ministry of Energy and Mines. *Decreto Supremo que aprueba el Reglamento de Seguridad y Salud Ocupacional y otras medidas complementarias en minería.* Decreto Supremo N.° 055-2010-EM de 01-01-2011, Lima, Peru, MINEM, 2011, available at [https://www.minem.gob.pe/minem/archivos/file/Mineria/LEGISLACION/2010/AGOSTO/DS%2005 5-2010--EM.pdf].

Source: Silvia Estefanía Lazo González. "Implementación de sistema de gestión de seguridad y salud en el trabajo para la realización de lasaña tradicional en la empresa Pizzerias Presto en base a Ley N.° 28783 Ley de Seguridad y Salud en el Trabajo" (Bachelor's thesis). Arequipa, Peru, Universidad Católica de Santa María, 2016, available at [https://repositorio.ucsm.edu.pe/handle/20.500.12920/5518].

CHAPTER TWO :

LEGAL FRAMEWORK ON OCCUPATIONAL RISK MANAGEMENT

I. Supreme Decrees related to occupational risk control

According to the MTPE[23] , it applies to all employees of the various economic sectors subject to the private labor regime (services, industry, education, fishing, clothing, etc.), not only to those with special rules on the subject, such as electricity or mining.

While in the supreme decree of the MVCS[24] , the "National Sanitation Plan 2017- 2021" is indicated. In Peru, the services offered to improve the sanitary conditions of citizens are inadequate because they lack quality and continuity, which indicates a lack of infrastructure to provide this type of services in the country. According to projections by the National Institute of Statistics and Informatics (INEI), in 2016 Peru had an estimated population of 32.4 million inhabitants, of which 72.2% corresponds to the urban area; while 22.8% to the rural area. The coverage estimates recorded indicate that in the urban area 94.5% of the total inhabitants have drinking water services and 88.3% have sewerage services. On the other hand, in rural areas, coverage is estimated at 71.2% for drinking water and 24.6% for sewerage[25] .

---

[23] Ministry of Labor and Employment Promotion. *Approval of occupational health and safety regulations*. Decreto Supremo N.° 009-2005-TR de 29-09-2005, Lima, Peru, MTPE, 2005, available at [https://oiss.org/wp-content/uploads/2018/11/12-03_Reglamento_de_Seguridad_y_salud_en_el_trabajo_2005-09-29_009-2005-TR_487.pdf].

[24] Ministry of Housing, Construction and Sanitation. *Decreto Supremo que modifica el Reglamento de Organización y Funciones del Ministerio de Vivienda, Construcción y Saneamiento*. Supreme Decree No. 010-2014-VIVIENDA of 03-03-2015, Lima, Peru, MVCS, 2015, available at [https://cdn.www.gob.pe/uploads/document/file/360774/DS-006-2015-VIVIENDA.pdf].

[25] National Institute of Statistics and Informatics. *Estado de la población peruana 2020*. Lima, Peru, INEI, 2020, available at [https://www.inei.gob.pe/media/MenuRecursivo/publicaciones_digitales/Est/Lib1743/Libro.pdf].

II. Occupational health and safety laws

According to Law No. 29783, which contains a new legal basis for preventing risks at work, it stipulates in its additional provision that it is important to adapt the rules of the ministries to increase the risk culture in a nation[26] .

And Law No. 30222[27] , which was published three years later with certain modifications to the aforementioned law, provides for certain general and specific rules for:

— To guarantee the psychophysical integrity of the people involved in the development of the activities, through the identification, recognition, reduction and control of risks in order to reduce the occurrence of incidents.

— Execute work functions in healthy and safe environments.

— To develop instructions for the execution of management plans, as well as to mitigate risk cases.

— Promote a culture of risk prevention in the workplace, in relation to the performance of their respective functions.

— Consider the inclusion of all workers in the occupational health and safety system.

Structural content of the occupational health and safety law

---

[26] Congress of the Republic. *Law of Safety and Health at Work*. Law No. 29783 of 20-08-2011. Lima, Perú, Congreso de la República, 2011, available at [https://web.ins.gob.pe/sites/default/files/Archivos/Ley%2029783%20SEGURIDAD%20SALUD%20EN%20EL%20TRABAJO.pdf].

[27] Congress of the Republic. *Law that modifies Law 29783, Law of Safety and Health at Work*. Law No. 30222 of 11-07-2014. Lima, Peru, Congreso de la República, 2014, available at [https://leyes.congreso.gob.pe/Documentos/Leyes/30222.pdf].

According to the MTPE[28] the only legal provision or TUO that allows regulating and preventing risks in companies is Law No. 29783.

Likewise, a system was established that allows the protection of an employee's health, consisting of national and regional councils[29] . The National Council is made up of 12 representatives, either from the State (4), the company itself (4) and the union (4).

Among the functions exercised by an employer is the registration of the OSHMS, which can be kept for 20 years if cases of occupational diseases are accounted for. If the company has 20 or more employees, a safety committee and occupational health and safety procedures are required; otherwise a supervisor will be assigned. Similarly, a minimum of four training sessions per year must be attended regarding occupational safety and health, design risk maps of the company, among others.

The employer's failure to comply with the duty of prevention generates the obligation to pay compensation to the victims or their beneficiaries.

On the other hand, the rights and obligations of the personnel include the following:

— They convey the actions and demands of employees.

— They are protected against acts of hostility from the employer.

— Participate in training programs.

— They are entitled to a suitable job.

— The protection extends to workers of contractors and subcontractors.

---

[28] Ministry of Labor and Employment Promotion. *Ley de Seguridad y Salud en el Trabajo, its regulations and amendments.* Lima, Peru, MTPE, 2017, available at [https://cdn.www.gob.pe/uploads/document/file/349382/LEY_DE_SEGURIDAD_Y_SALUD_EN_E L_TRABAJO.pdf].

[29] Congress of the Republic. *Law on Safety and Health at Work*, cit.

— Obligations are established that employees must comply with (e.g., comply with rules and regulations, use assigned instruments and work materials, not manipulate equipment and tools without authorization, cooperate in investigation processes, submit to examinations, inform the employer of any risk event, report accidents, etc.).

A new article (article 168) is also incorporated regarding healthy and safe working conditions.

In addition, this law is regulated by Supreme Decree 005-2012-TR[30] , made up of 123 articles, which are organized in seven titles, a final and transitory provision.

Table 1. Structure of Supreme Decree 005-2012-TR

Title	Chapter	Articles
I.- General Provisions	General	1 to 4
II.- National OSH Policy		5 to 6
III.- National OSH system	I	7 to 21
	II	22
IV.- OSH Management System	I	23 to 24
	II	25
	III	26 to 37
	IV	38 to 73
	V	74 to 75
	VI	76 to 78
	VII	79 to 84
	VIII	85 to 88

---

[30] Presidency of the Republic of Peru. *Approve the Law on Safety and Health at Work*. Supreme Decree No. 005-2012-TR of 01-11-2016. Lima, Peru, Presidency of the Republic of Peru, 2016, available at [https://www.gob.pe/institucion/presidencia/normas-legales/462577-005-2012-tr].

	IX	89 to 90
V.-Rights and obligations	I	92 to 104
	II	105 to 109
VI.- Notification of accidents		110 to 116
	I	117 to 118
	II	119 to 122
VII.- Supervision and control		123
Final Supplementary Provision		Only
Supplementary transitory provisions		14

Source: Presidency of the Republic of Peru. *Approval of the Occupational Health and Safety Law*, cit.

Supreme Decree 005-2012-TR clearly states that it is necessary to include the company's health and safety protocols in the labor contract.

These must take into account occupational risks, specifically those related to the functions performed in the company; it is not only intended to provide informative brochures, a corporate commitment is required so that the personnel can easily identify and understand the risks.

Paid leave will be granted for the time used to travel to the training site, the time spent at the training site and the time required to return to the work center.

III. OHSAS 18001 Standard

This standard is used globally for the occupational risk management system (OHSMS), in place of other models that have already disappeared, such as the BS 8800 Guide. It also has ISO 9001 and ISO 14001, prestigious standardization and certification bodies, in addition to the ILO approach which, although not defined as "non-certifiable", does not promote certification[31] .

A. Structure of the OHSAS standard 18001

OHSAS 18001, Occupational Health and Safety Assessment Series, refers to a globally recognized assessment standard for OHSMS, developed by a group of leading trade and certification organizations to fill the niche in international standards[32] .

This standard considers the following basic areas:

— Identifying threats, assessing risks and establishing controls
— Legal and other requirements
— Objectives and programs
— Resources, positions, responsibility, duty and authority
— Competence, training and awareness
— Communication, participation and consulting
— Operational control.
— Emergency preparedness and response.
— Performance measurement, monitoring and improvement

Likewise, the OSHMS model proposed by this standard is structured in five modules:

— OSH Policy
— Planning
— Implementation and operation
— Verification
— Management review

Figure 3. OHSAS 18001 standard model

---

[31] Spanish Association for Standardization and Certification (Ed.). *OHSAS 18001:2007. Occupational health and safety management systems - Requirements*, cit.
[32] Idem.

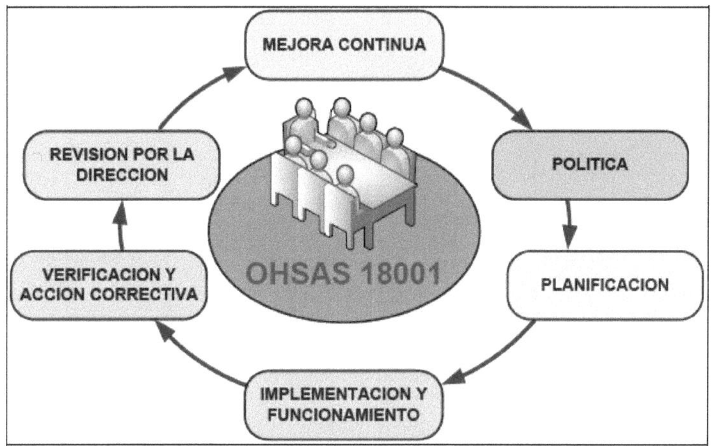

Source: Spanish Association for Standardization and Certification (Ed.). *OHSAS 18001:2007. Occupational health and safety management systems - Requirements*, cit.

There are also other advantages for proper business management within the health and safety framework of this regulation.

— Enables the integration of health and safety by levels.

— It fosters positive worker attitudes, motivates continuous participation and promotes a culture of prevention in a structured environment.

— Facilitates tools to reduce work-related incidents.

— It allows to comply and demonstrate compliance with legality.

— It makes the company's image stand out and present itself to society.

This standard is more compatible with ISO 14001 and ISO 9001, encompasses modern and proven concepts regarding occupational health and safety management, as well as other meanings[33] .

Figure 4. Compatibility between regulations

---

[33] Cristina Abril Sánchez, Antonio Enríquez Palomino and José Manuel Sánchez Rivero. *Guía para la integración de sistemas de gestión: calidad, medio ambiente y salud en el trabajo*, 2nd ed., Madrid, Spain, Fundación Confemetal, 2012.

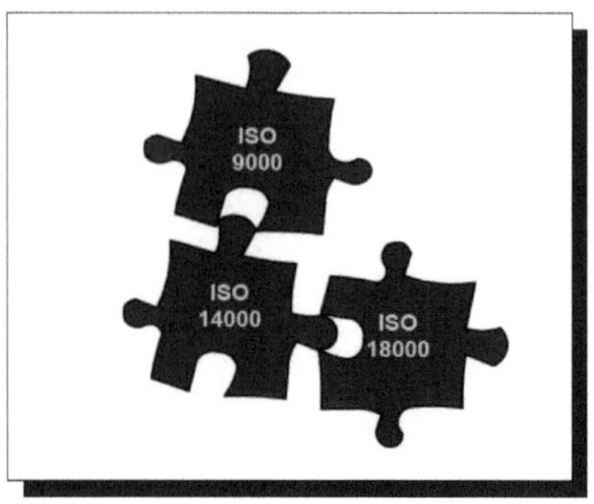

Source: Carlos Fernando Atahualpa Carrera Endara, Cristian Heriberto Ligña Cumbal, Galo Renan Moreno Cueva and Ruben Morales Carrera. *Sistemas de gestión de calidad.* United States, Compás, 2018, available at [http://142.93.18.15:8080/jspui/bitstream/123456789/466/3/SISTEMAS%20DE%20GESTI%C3%93N%20DE%20LA%20CALIDAD.pdf].

It should be noted that the OHSAS 18001 standard is based on the PHVA methodology, also known as Edward Deming's cycle[34] .

Figure 5. Deming Cycle

---

[34] Carrera Endara, Ligña Cumbal, Moreno Cueva and Morales Carrera. *Sistemas de gestión de calidad,* cit.

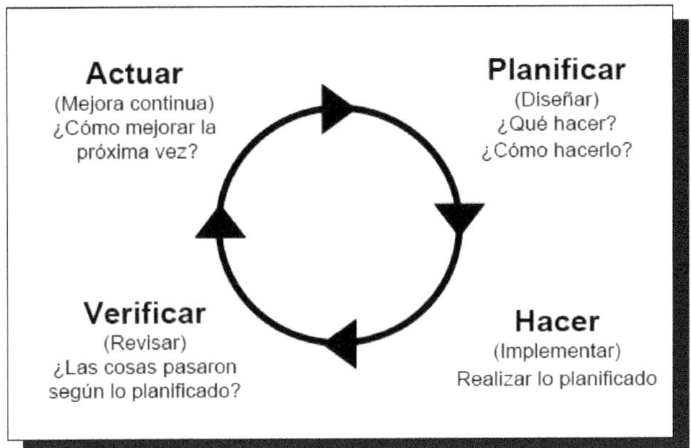

Source: Carrera Endara, Ligña Cumbal, Moreno Cueva and Morales Carrera. *Sistemas de gestión de calidad*, cit.

The structure of the model includes, among others, general requirements, objectives and programs, IPERC, communication, documentation, operational control, legal compliance assessment, accident investigation, nonconformities, records control, internal audit and management review[35] .

Table 2. Structure of the OHSAS 18001 standard

---

[35] Bernal Mateus, María del Carmen. *La norma OHSAS 18001 y su implementación*, 2nd ed., Bogotá, Colombia, ICONTEC, 2009.

<table>
<tr><td colspan="2">Structure of the OHSAS 18001 Standard</td></tr>
<tr><td colspan="2">

1. PURPOSE AND SCOPE OF APPLICATION

2. PUBLICATION FOR CONSULTATION

3. TERMS AND DEFINITIONS

4. REQUIREMENTS OF THE SST MANAGEMENT SYSTEM:

    4.1   General requirements

    4.2   OSH Policy

    4.3   Planning

        4.3.1   Hazard identification, risk assessment and controls

        4.3.2   Legal and other requirements

        4.3.3   Objectives and programs

    4.4 Implementation and operation

        4.4.1   Resources, roles, responsibilities and authority

        4.4.2   Competence, training and awareness

        4.4.3   Communication, participation and consultation

        4.4.4   Documentation

        4.4.5   Document control

        4.4.6   Operation control

        4.4.7   Emergency preparedness and response

</td></tr>
<tr><td colspan="2">

    4.5   Verification

        4.5.1   Performance monitoring and measurement

        4.5.2   Legal compliance assessment

        4.5.3   Incident investigation, nonconformity, CA/AP

        4.5.4   Control of records

        4.5.5   Internal audit

    4.6 Management review.

</td></tr>
</table>

Source: Spanish Association for Standardization and Certification (Ed.). *OHSAS 18001:2007. Occupational health and safety management systems - Requirements*, cit.

The OHSAS 18001 standard cycle is influenced by business needs to increase competitiveness, considering the losses arising from injuries suffered by its workers.

Figure 6. The OHSAS 18001 system cycle

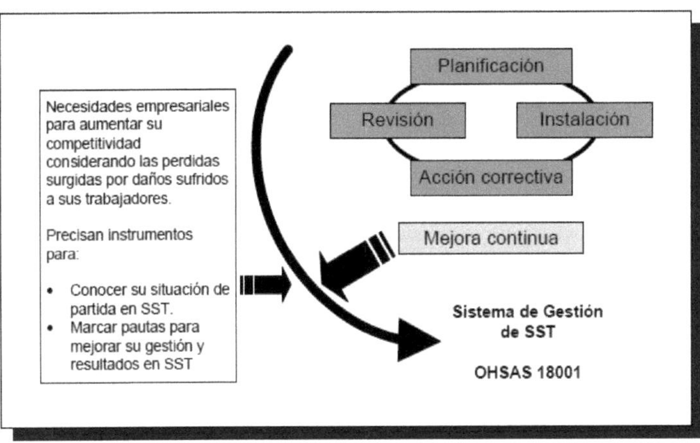

Source: Spanish Association for Standardization and Certification (Ed.). *OHSAS 18001:2007. Occupational health and safety management systems - Requirements*, cit.

The following figure shows the elements involved in the OSHMS:

Figure 7. Elements involved in the OSHMS and their functions.

Source: Spanish Association for Standardization and Certification (Ed.). *OHSAS 18001:2007. Occupational health and safety management systems - Requirements*, cit.

B. Documentation

This standard stipulates the requirements for an OSHMS, which enables a company to manage occupational risks and improve its OSH performance, but does not specify OSH performance criteria or give detailed specifications for the design of an OSHMS. The design must be elaborated by the organization, in this case by the company, according to its size, risks, type, activities and resources. This elaboration is done by identifying specific procedures and requires the following documents:

— Occupational Safety and Health (OSH) Manual; set of definitions that establish the actions and procedures for planning, operation, verification and revision of the standard.

— Administrative management procedures; an orderly set of instructions or rules that define and establish the operating standards required by the standard and applied to the company.

— Procedures for critical activities; specific instructions or rules that are defined for the control of risks in activities defined as critical.

— Record control procedures; An orderly set of instructions that establishes the accreditation of the records required by the standard for the continuous verification of compliance.

Figure 8. Documentation in the OHSAS 18001 system

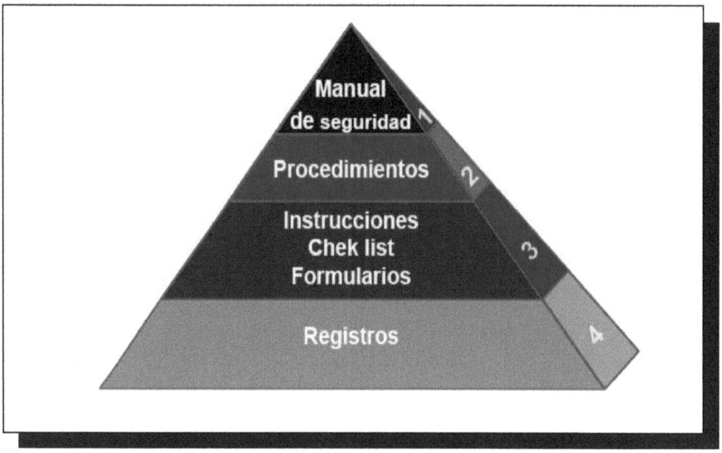

Source: Spanish Association for Standardization and Certification (Ed.). *OHSAS 18001:2007. Occupational health and safety management systems - Requirements*, cit.

C. Benefits of implementation

One of the benefits of this standard is that it will reduce the number of personnel involved in accidents (own and third parties) through prevention and risk management in the company; as well as the resources wasted due to accidents. Similarly, customers will be satisfied with the service or product offered. The company's stakeholders will be able to see that there is a commitment to the health and safety of the personnel.

In addition, it must be ensured that the respective legislation is complied with; the non-existence of unwanted stoppages, with the consequent reduction of costs. In short, productivity and therefore competitiveness will be improved.

CHAPTER THREE :

PRELIMINARY STUDY

This chapter will present various studies carried out on the application of standards to control risks in different organizations, and will also consider the requirements and regulations of the companies themselves to prevent risks. Finally, the situation of Peruvian companies on this topic is detailed.

As expressed by Pérez[36] in his research, all companies must have a system that allows them to provide safety and health to their personnel, which will be necessary to establish the guidelines for an adequate organizational management. Therefore, by implementing such a system in the contractor companies of the mining-metallurgical economic sector, the trend of fatal accidents will be reduced.

González[37] developed a system based on the OHSAS 18001 standard to reduce the risks to which workers in a cosmetics factory were exposed. This was done to contribute to the well-being of the employees and thus increase their productivity in the factory.

As for the Valladarez study[38] , it established a structure referring to the recognition and control of risks to reduce the possibility of accidents, this also

---

[36] José Luis Pérez. "Sistema de gestión en seguridad y salud ocupacional aplicado a empresas contratistas en el sector económico minero metalúrgico" (Master's thesis). Lima, Perú, Universidad Nacional de Ingeniería, 2007, available at [http://hdl.handle.net/20.500.14076/633].

[37] Nury Amparo González González. "Diseño del sistema de gestión en seguridad y salud ocupacional, bajo los requisitos de la norma NTC-OHSAS 18001 en el proceso de fabricación de cosméticos para la empresa Wilcos S.A." (bachelor's thesis). Bogotá, Colombia, Pontificia Universidad Javeriana, 2009, available at [http://hdl.handle.net/10554/7232].

[38] Jaime Mauricio Valladarez Tola. "Implementación del sistema de gestión en seguridad y salud ocupacional bajo una nueva versión de la norma OHSAS 18001:2007 en la Corporación Eléctrica de Ecuador Celec-Hidropaute" (Master's thesis). Ecuador, Universidad de Cuenca, 2010, available at [http://dspace.ucuenca.edu.ec/handle/123456789/2631].

contributed to the improvement of the general performance of workers, under the new version of the OHSAS 18001 standard in the electrical corporation of Ecuador.

For his part, Terán[39] proposed to achieve a more effective performance in the field of prevention, through a process of continuous improvement. Thus, companies can also have at their disposal an essential tool for carrying out the stipulations of the OHSAS 18001 standard.

Regarding Matabanchoy's research[40] , it was executed from the choice and appropriate distribution of several digital documents. The challenge has been to promote a healthy organization that ensures well-being and an adequate working environment, in which it is essential to have the support of senior management.

While in the study by Bustamante[41] , it is inferred that there was an effective responsibility of the presidency of the electrical construction company IELCO to periodically monitor compliance with safety regulations within the administrative and project execution areas. It should also be considered that this company was able to design proposals based on continuous improvement to safeguard worker health. This has been the way to start the process of restructuring a company to implement the normative standard (OHSAS 18001).

In relation to Romero's research[42] , an analysis was made of the problems encountered at Mirrorteck Industries S.A., which does not have an occupational health

---

[39] Itala Sabrina Terán Pareja. "Propuesta de implementación de un sistema de gestión de seguridad y salud ocupacional bajo la norma OHSAS 18001 en una Empresa de Capacitación Técnica para la Industria" (Bachelor's thesis). Lima, Pontificia Universidad Católica del Perú, 2012, available at [http://hdl.handle.net/20.500.12404/1620].

[40] Sonia Maritza Matabanchoy Tulcán. "Salud en el trabajo." *Universidad y Salud*, vol. 14, no. 1, 2012, pp. 87-102, available at [https://revistas.udenar.edu.co/index.php/usalud/article/view/1270].

[41] Fernando Bustamante Granda. "Sistema de gestión en seguridad basado en la norma OHSAS 18001 para la empresa constructora eléctrica IELCO" (Master's thesis). Ecuador, Universidad Politécnica Salesiana, 2013, available at [https://dspace.ups.edu.ec/handle/123456789/5375].

[42] Ángela Liliana Romero Albán. "Diagnóstico de normas de seguridad y salud en el trabajo e implementación del reglamento de seguridad y salud en el trabajo en la Empresa Mirrorteck Industries S.A." (Master's thesis). Ecuador, Universidad de Guayaquil, 2014, available at [http://repositorio.ug.edu.ec/handle/redug/4494].

and safety management system model, as required by Ecuadorian legislation. The methodology used is reflective and descriptive, it also analyzes the problems, evaluates the costs and proposes solutions and training to the plant personnel.

Vásquez's research[43] , which contributes to the continuous improvement of the organization through the integration of prevention at all hierarchical levels of the company and the use of improvement tools, indicates that 23 potential hazards were identified and that a total of 132 workers were exposed. The risks were considered moderate and occurred mostly in drilling, mechanical equipment and transportation activities, while there were fewer in monitoring and office activities.

On the other hand, Achinte and Henao[44] made a diagnosis of the current state of a maintenance company, then a plan was implemented to collect and record the information required for planning, and finally a proposal was developed on how to structure the occupational health and safety system. The organization had data referring to the aforementioned, however, it was in a disorganized manner and prevented the execution of the normative guidelines. It was concluded that 80% of the planned system was implemented.

Regarding the study by Barrera-García *et al.*[45] , the application of mathematical models to prevent occupational injuries is proposed. This should take into account the type of task and its characteristics, the work system, the technical design, the safe environment within the company and the risk culture, as well as the characteristics of

---

[43] Marco Antonio Vásquez Ojeda. "Implantación de un sistema de gestión de seguridad y salud ocupacional en el proyecto especial Olmos - Tinajones - Lambayeque" (Master's thesis). Trujillo, Peru, Universidad de Trujillo, 2016, available [http://dspace.unitru.edu.pe/items/62188267-261a-49a1-bf9d-9ad96b750193].

[44] Adriana Stella Achinte Hurtado and Sidney Oriana Henao Clavijo. "Planning of the occupational health and safety management system for a locative maintenance company based on Decree 1072 of 2015, period 2015-2016" (specialty thesis). Cali, Colombia, Universidad Libre, 2016, available at [https://repository.unilibre.edu.co/handle/10901/9893?show=full].

[45] Aníbal Barrera-García, Alejandro González-Delgado and Damayse Pérez-Fernández. "Identification of incident factors in occupational accident rates in companies in Cienfuegos." *Industrial Engineering*, vol. 37, no. 2, 2016, pp. 127-137, available at [https://www.redalyc.org/articulo.oa?id=360446197003].

the employee. All these elements can have an impact on occupational safety, the work environment, safe behaviors and, consequently, on the reduction of injuries.

According to Gadea[46] , the inappropriate handling of cargo, the continuous entry of trucks and storage of harmful products, the lack of adequate lighting are some factors that may have an impact on the increase of risks within a company; however, the most critical are those made directly with the production, which in this case refers to the operations with the cutting of fabric and sewing. Therefore, it is necessary to apply an OSHMS to increase the administrative, ethical and industrial benefits in the company.

Purga and Torres[47] point out that their plan to implement a system to ensure the health and safety of an employee in the work environment allows for a decrease in costs for respective regulatory violations. However, the decrease for accidents at work is not feasible since the obligation of employees to comply with safety systems exceeds the management parameters.

As for the study by Tirado and Vega[48] , an occupational risk plan was designed for a company that provides drinking water services, which contains the company's regulations, an IPERC matrix, annual programs related to personnel welfare and written procedures for safe work. Finally, an economic evaluation of the proposed plan has been carried out and the economic indicators obtained are NPV = S/. 140,384.41

---

[46] Adrián Wilfredo Gadea García. "Propuesta para la implementación del sistema de gestión de seguridad y salud en el trabajo en la empresa SUMIT S.A.C." (Bachelor's thesis). Lima, Peru, Universidad de Lima, 2016, available at [https://repositorio.ulima.edu.pe/bitstream/handle/20.500.12724/3497/Gadea_Garcia_Adrian.pdf?sequence=1&isAllowed=y].

[47] Wendy Arelí Purga Ruiz and Anthony Percy Torres Vargas. "Propuesta de implementación de un sistema de seguridad y salud ocupacional basado en la norma OHSAS 18001:2007 para evitar costos por incidentes en el consorcio Alvac Johesa" (undergraduate thesis). Lima, Peru, Universidad Privada del Norte, 2017, available at [https://repositorio.upn.edu.pe/handle/11537/12390].

[48] Jefferson Andrée Tirado Medina and Víctor Luis Vega Ybáñez. "Propuesta para la implementación de un plan de seguridad y salud ocupacional para controlar los riesgos y reducir los accidentes en la división de mantenimiento de la empresa de servicio de agua potable y alcantarillado de la Libertad - Sedalib S.A." (Bachelor's thesis). Trujillo, Peru, Universidad Nacional de Trujillo, 2017, available at [http://dspace.unitru.edu.pe/handle/UNITRU/8880].

and IRR = 72%, which means that the implementation of the proposal has been highly profitable.

On the other hand, Niciejewska and Kiriliuk[49] consider that the main concern that a company should have is the ability to classify and assign appropriately the potential risks of the personnel in its company, evaluate them and take the necessary corrective or preventive measures.

Regarding the research by Contri and Desiderio[50] , in which an OSHMS based on the ISO 45001 standard was applied, it is stated that this is an opportunity for large companies to work on various aspects of OSH through a highly competitive management approach, since it enables the consolidation of an organizational culture, the participation of employees and the commitment of top management, improving personnel training and managing risk prevention through action plans.

For their part, Cuba and Mercado[51] consider that personnel must be trained not only in occupational hazards, but also in the application of work tools to carry out activities effectively and act quickly in case of emergencies.

Peruvian occupational health and safety situation

Mejía *et al.*[52] conducted an investigation on the trend of occupational accidents and diseases reported to the Peruvian Ministry of Labor and Employment Promotion

---

[49] Marta Niciejewska and Olga Kiriliuk. "Occupational health and safety management in 'small size' enterprises, with particular emphasis on hazards identification." *Production Engineering Archives*, vol. 26, no. 4, 2020, pp. 195-201, available at [https://sciendo.com/article/10.30657/pea.2020.26.34].

[50] Leandro Contri Campanelli and Lucas Desiderio Ribeiro. "Involvement of Brazilian companies with occupational health and safety aspects and the new ISO 45001:2018." *Production*, vol. 31, 2021, pp. 1-13, available at [https://doi.org/10.1590/0103-6513.20210005].

[51] Ramiro Cuba Miranda and César Mercado Rivero. "Implementación de un plan de seguridad y salud ocupacional en las labores de mantenimiento, planchado y pintura en la empresa Fátima Car Service Srl - Cusco - 2021" (Bachelor's thesis). Cusco, Peru, Universidad Continental, 2022, available at [https://hdl.handle.net/20.500.12394/11814].

[52] Christian Mejía, Matlin Cárdenas and Raúl Gomero-Cuadra. "Notification of occupational accidents and diseases to the Ministry of Labor. Peru 2010-2014." *Revista Peruana de Medicina Experimental y Salud Pública*, vol. 32, no. 3, 2015, pp. 526-531, available at [https://doi.org/10.17843/rpmesp.2015.323.1689].

(MTPE). A descriptive study was carried out with reports extracted from monthly bulletins from September 2010 to December 2014, nationwide (54 596) non-fatal occupational accidents were reported, of which 90.2% (48 365) were male. Metropolitan Lima was the city where the highest number of non-fatal occupational accidents were reported (76.9%), followed by the constitutional province of Callao (15.0%) and the department of Arequipa (3.8%). During the same period, 674 fatal accidents, 3432 incidents and 346 occupational diseases were reported.

Table 3. Occupational illnesses, accidents and incidents reported to the MTPE from 2010 to 2014.

Notifications	2010	2011	2012	2013	2014	Total
Non-fatal occupational accidents	198	4 728	15 508	19 412	11 271	54 596
Metropolitan Lima	140	4117	11 630	14 804	14 570	41 962
Callao	0	301	3430	3481	991	8203
Arequipa	0	29	183	222	1647	2 081
Piura	14	100	410	531	413	1468
La Libertad	0	20	79	101	87	287
Fatal occupational accidents	24	145	199	178	128	674

Labor incidents	130	623	826	983	826	2432
Diseases labor	8	110	107	32	39	346

Source: Mejía, Cárdenas and Gomero-Cuadra. "Notification of occupational accidents and diseases to the Ministry of Labor. Peru 2010-2014," cit.

Table 4. Types of occupational diseases reported to the MTPE from 2010 to 2014.

Occupational diseases	2010	2011	2012	2013	2014	Total
Hearing loss	0	28	21	24	4	77
Allergic dermatitis	0	3	30	7	4	44
Silicosis	0	9	6	18	4	37
By positions	0	38	4	15	0	57
Leishmaniasis	0	7	5	2	3	17
Due to toxics/chemicals	0	9	5	0	2	16
Dorsalgia	0	0	3	0	0	3
Varicose veins in lower limbs	0	0	0	1	0	1
Sciatica	0	0	1	0	0	1

| Due to radiation | 0 | 1 | 1 | 0 | 0 | 2 |
| Hepatitis | 0 | 0 | 1 | 0 | 0 | 1 |

Source: Mejía, Cárdenas and Gomero-Cuadra. "Notification of occupational accidents and diseases to the Ministry of Labor. Peru 2010-2014," cit.

Figure 9. Non-fatal occupational accidents reported from 2010 to 2014.

Source: Mejía, Cárdenas and Gomero-Cuadra. "Notification of occupational accidents and diseases to the Ministry of Labor. Peru 2010-2014," cit.

Figure 9 shows the variation of lethal accidents from 2010 to 2014. It is observed that in 2013 there were 19,412 non-fatal accidents at national level, which has decreased in 2014 to 14,804 non-fatal accidents.

Figure 10. Lethal occupational accidents reported from 2010 to 2014.

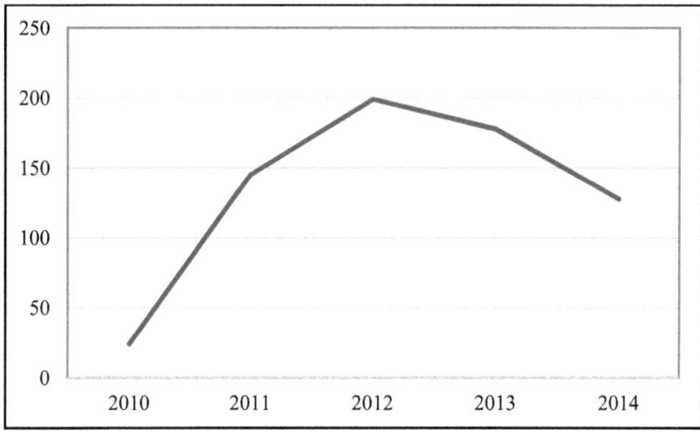

Source: Mejía, Cárdenas and Gomero-Cuadra. "Notification of occupational accidents and diseases to the Ministry of Labor. Peru 2010-2014," cit.

Figure 10 shows the variation from 2010 to 2014. It is noted that in 2012 there were 199 fatal accidents at national level, decreasing in 2014 to 128 fatal accidents.

Figure 11. Occupational incidents from 2010 to 2014.

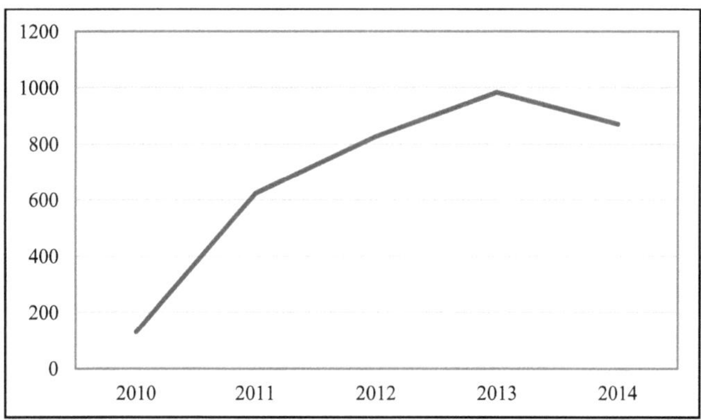

Source: Mejía, Cárdenas and Gomero-Cuadra. "Notification of occupational accidents and diseases to the Ministry of Labor. Peru 2010-2014," cit.

Figure 11 shows the variation from 2010 to 2014. In this regard, in 2013 there were 983 incidents at the national level, this has decreased in 2014 to 870 incidents.

Companies and the State must implement strategies to reduce risk by providing training to workers, safety material and inspections by the National Superintendence of Labor Inspection (SUNAFIL, 2016).

Figure 12. Occupational illnesses from year 2010 to year 2014.

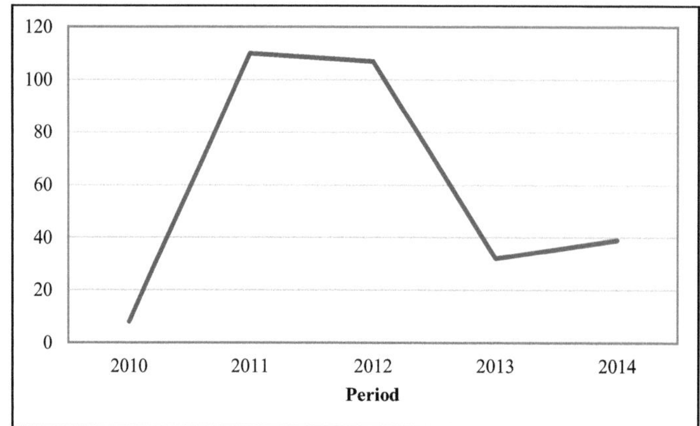

Source. Mejía, Cárdenas and Gomero-Cuadra. "Notification of occupational accidents and diseases to the Ministry of Labor. Peru 2010-2014," cit.

Figure 12 shows the variation from 2010 to 2014. Likewise, it can be seen that in 2011 and 2012 there was the highest number of occupational diseases at national level, which decreased in 2014.

CHAPTER FOUR :

PROPOSAL FOR PLANNING AN OCCUPATIONAL HEALTH AND SAFETY SYSTEM FOR A SANITATION COMPANY. CASE STUDY

I. Problem statement

Corporate security is that which deals with rules, procedures and strategies aimed at preserving the physical integrity of personnel[53] .

Therefore, every existing organization must be a benchmark for labor sustainability. For this reason, they cannot continue only with the production of goods and services, but must be able, on a daily basis, to take on the challenge of reducing their impact on occupational risks, for the benefit of their environment.

According to statistical data from Peru's MTPE, between the years 2010 and 2014, 54 596 non-lethal occupational accidents and 41 962 lethal occupational accidents were reported nationwide[54] . This shows that there are a large number of accidents occurring in the country's workplaces.

SEDA-JULIACA, which provides drinking water and sewage services to various sectors of Peru, complies with health standards. Similarly, the main activities are carried out in plants that treat wastewater, production area, and include the following operations: cleaning catchment area of the Coata River catchment area, Ayabacas sector; pumping and impulsion; equipment maintenance (electric pumps); sedimentation and cleaning of settling tanks; and cleaning of conventional filters. However, it was found that the workers do not have adequate PPE (personal protective

---

[53] Congress of the Republic. *Law on Occupational Safety and Health*, cit.
[54] Mejía, Cárdenas and Gomero-Cuadra. "Notification of occupational accidents and diseases to the Ministry of Labor. Peru 2010-2014," cit.

equipment) for each of the activities, nor are there any safety signs; therefore, the workers are exposed to hazards.

Faced with this work situation, an OSHMS should be implemented to reduce accidents in the workplace, reduce downtime and associated costs, and clarify your commitment to employee health and safety.

Standardized management systems are proven management and documentation tools based on "standardized" procedures and are a good tool to prevent accidents in companies.

Therefore, it is proposed to design an OSHMS according to the OHSAS 18001 standard for the company SEDA-JULIACA, based on a "systemic diagnosis of risk prevention"[55] from the organizational point of view, which includes the most evident risk effects. That is, by identifying the problems and diagnosing the functioning of the health and safety management system in the company.

II. Problem of the study

A. General problem

How would the proposed planning of an Occupational Health and Safety Management System benefit the sanitation company SEDA-JULIACA according to the OHSAS 18001 standard?

B. Specific problems

- How to make a detailed diagnosis of the current situation of SGSST of the company SEDA-JULIACA?

---

[55] Morell González, Luisa María; Rosa Maricela Cedeño Zambrano and Sarai Ramírez Cruz. "Systemic management of institutional risks. Diagnosis in the Agrarian University of Havana and the Technical University of Manabí". *Cofin Habana*, vol. 13, no. 1, 2019, pp. 1-10, available at [http://scielo.sld.cu/scielo.php?script=sci_arttext&pid=S2073-60612019000300016], p. 1.

- What aspects should be included in the Occupational Health and Safety Management System based on the OHSAS 18001 standard proposed for SEDA-JULIACA?

- Should a hazard identification and risk assessment matrix be developed?

III. Objectives of the study

A. General Objective

To propose the planning of an Occupational Health and Safety Management System for the sanitation company SEDA-JULIACA based on the international standard OHSAS 18001.

B. Specific objectives

- Diagnose the OSHMS situation of the sanitation company SEDA-JULIACA.

- Determine what aspects the occupational health and safety management system should contain in the SEDA-JULIACA company.

- Elaborate hazard identification and risk assessment matrix (IPERC).

IV. Study hypothesis

A. General hypothesis

The proposal for planning an OSHMS, based on the OHSAS 18001 standard, in the sanitation company SEDA-JULIACA will improve the occupational health and safety management system.

B. Specific hypotheses

- A diagnosis will be made of the situation of the SGSST of the SEDA-JULIACA company.

- The aspects that must be included in the occupational health and safety management system in the sanitation company SEDA-JULIACA will be determined.

- A hazard identification and risk assessment matrix (IPERC) will be developed.

V. Justification

In the first instance, it is considered that the proposal integrates actions that contribute to the solution of diagnosed problems and a set of parameters that allow their measurement, control and follow-up. The clear determination of the process by the top management of the company SEDA-JULIACA is an essential element in its development.

Therefore, carrying out an OSHMS diagnosis in this company will allow us to have an understanding of:

— The status of OSHMS in the company, from which a correct policy and objectives of the safety system can be defined to make possible the development of the implementation of the OSHMS.
— The identification of those risk incidents that affect the company, with the purpose of correcting them.
— Comply with national occupational health and safety regulations.
— To implement the system based on the OHSAS 18001 standard, so that company personnel can easily participate in it.

In addition, this proposal will make it possible to reduce the number of incidents and non-productive periods, the company's commitment to provide safety and health to workers, strengthen resources in a sustainable and accessible environment, so as to contribute to the management of human resources for the development of society.

VI. Variables of study

The hypothesis variables are as follows:

— Independent variable: international standard OHSAS 18001.

— Dependent variable: planning of a Health and Safety Management System.

— Intervening Variable: SEDA-JULIACA's occupational health and safety diagnosis.

VII. Population and sample

The study population is composed of 175 workers who work in the sanitation company SEDA JULIACA, of which 6 are trusted officials, 63 are permanent workers (employees and workers), 33 are contracted workers (employees and workers) and 73 are temporary or non-personal service personnel.

The sample considered for this study is composed of 30 workers at the treatment plant. A non-probabilistic sample has been taken, since it depends on "causes related to the characteristics of the research, i.e., the sample obeys the importance of resource management"[56] .

Study area

This research was conducted in the company SEDA-JULIACA, specifically in the area where the drinking water treatment plant is located. It is located in the city of Juliaca, province of San Román, department of Puno.

Figure 13. Satellite visualization of the company SEDA-JULIACA.

---

[56] Roberto Hernández-Sampieri and Christian Paulina Mendoza Torres. *Research methodology: the quantitative, qualitative and mixed routes*. Mexico City, McGraw-Hill Interamericana Editores, 2018, p. 200.

Source: EPS SEDA-JULIACA. *General company data*. Juliaca, Peru, EPES SEDA-JULIACA, 2023, available at [https://SEDA-JULIACA.com/datos-dela-empresa/].

Figure 14. Drinking water treatment plant area.

Source: EPS SEDA-JULIACA. *General company data*. Juliaca, cit.

VIII. Type and focus of study

This study corresponds to a qualitative approach, of an explanatory descriptive type, since it details particular situations, events and facts that have occurred. Descriptive studies seek to specify the properties, characteristics and profiles of people and communities, in this case of an organization under analysis. While the applicative studies seek that all the points developed in the research be applied in the activities developed.

IX. Data collection techniques and instruments

A bibliographic review of the norms applied to companies with respect to health and safety was carried out, as well as verification of the documentation concerning the company under study. For this reason, a preliminary diagnosis of the company SEDA-JULIACA and its risks was carried out.

In addition, interviews were conducted with employees of the company's treatment plant.

X. Data processing: prior diagnosis of the business situation

A. Preliminary diagnosis of the company

Description of the company

The company called EPS SEDA-JULIACA-S.A. is in charge of providing drinking water and sewage services in Juliaca. It is located more than 3,286 meters above sea level. The urban area extends over an area of approximately $41km^2$, and the topography is flat with a slope of 0.45 to 0.50 per meter.

Treatment plant

Juliaca has a surface water supply source that originates in the Coata River, whose waters are captured in the Ayabacas sector. To date, the treatment flow rate is between 300 L/s and 400 L/s. There are two treatment systems: a conventional one that treats 100 L/s, with a coarse sedimentation unit, a horizontal screen-type flocculator, and three fine sedimentation units; it is in fair condition due to repairs. The other system consists of two compact DEGREMON type units (sludge blanket), theoretically designed for a capacity of 120 l/s each.

Operations performed at the treatment plant

- *Catchment.* Water is pumped from the Coata River through five 10" and 14" diameter pipes, with an estimated capacity of 400 l/s. The water is collected directly into two cisterns, each installed in a horizontal pumping sump. The water is collected directly into two cisterns, each installed in a horizontal pumping sump; the collected water is then pumped to the treatment plant through a 24" diameter discharge pipeline at a distance of 85 m.
- *Sedimentation.* Treatment that consists of retaining the coarse particles by decantation, which sediment by their own weight or gravity.
- *Flocculation.* It intervenes in the hydraulic treatment unit of horizontal screens, in order to perform a slow agitation to form flocs, which is formed by the addition of a chemical component such as $Al_2(SO_4)3.5H\,O_2$
- *Filtration.* It is filtered with a battery of 10 sand filters, with a filtration capacity of 300 L/s and another 35 L/s pressure filter. The filters retain suspended particles that pass through the sedimentation process.
- *Disinfection.* There is a 500 Lb/24 hour chlorine dispenser.
- *Impulsion.* It has three horizontal pumping units of 100 L/s each and a vertical pumping unit with a flow of 50 L/s, which supply the city's reservoirs: Cerros Colorado and Santa Cruz, as well as Independencia; with a storage capacity of 10,745 m.[3]

Drinking water service and distribution

This company has more than 400 km of network ranging from 2" to 24" in diameter, made of cast iron, asbestos cement and PVC pipes. The potable water service provided to the population is 23,000 m^3 per day, with a coverage of 63% in an average of 5 hours per day, with estimated losses of approximately 30%.

House connections

The company has 41,314 total potable water connections, the percentage of micro-metering is quite low.

Quality control

This specialized laboratory is used to perform physicochemical and bacteriological analyses to ensure the quality of the drinking water provided to users.

Sewerage for wastewater treatment

The main chamber completely processes the inhabitants' wastewater, pumping 12,000 m^3 per day (62% coverage). In addition, seven auxiliary chambers have been installed in the area.

The wastewater treatment plant is located in the southwestern sector of the city, in the Chilla sector. It consists of eight oxidation ponds (primary facultative), which to date are not fulfilling their purpose due to the lack of treatment and removal of large quantities of sediment sludge. The effluent consists of a 21" diameter AC pipe that flows into the Torococha River.

Organization

The sanitation company SEDA-JULIACA has the following structure:

1. General Shareholders' Meeting

   — Directory
   — General Management

2. Control bodies

   — Institutional Control Office

— Advisory bodies

— Planning Office

— Office of Legal Counsel

— Institutional image office

— IT Office

Supporting bodies

— Administration Management

— Accounting Division

— Treasury Division

— Human Resources Division

— Supply Division

4. Line bodies

— Engineering Management

— Operations management

— Commercial Management

Figure 15. General organization chart of EPS SEDA-JULIACA S.A.

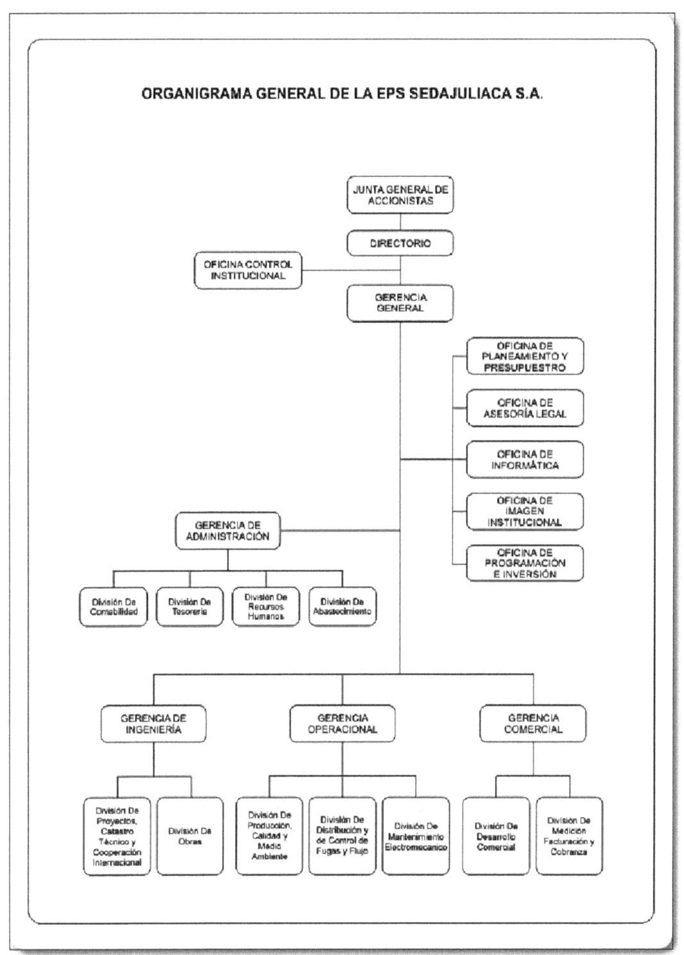

Source: EPS SEDA-JULIACA. *General company data*, cit.

Despite SEDA-JULIACA's extensive experience in providing potable water services, as well as the capacity of its workforce, it does not have any management system implemented in its organization, which is necessary to achieve the highest standards in occupational health and safety.

Thus, it is important that companies take into account Law No. 29783[57] , which regulates internal regulations in the framework of occupational health and safety.

Axiological principles

Every company must establish its axiological principles for a harmonious coexistence, SADAJULIACA is no exception, so its quality management contains several values, being these of knowledge of all its workers:

— vocation of service, being the user, the center of work, the responsibility of the company;

— respect;

— discipline, in terms of full compliance with the procedures and parameters established for production;

— occupational safety and

— respect for the environment.

Services provided

The sanitation company SEDA-JULIACA provides drinking water and sewerage services to 199,800 inhabitants of the province of San Román, in Juliaca.

The company administers and manages the economic resources to cover operating and maintenance costs through the commercial management.

Table 5. Users of SEDA-JULIACA

Services	Users (house connections)	% of coverage
Drinking water	49 940	75.50%

---

[57] Congress of the Republic. *Law on Occupational Safety and Health*, cit.

Sewer	50 925	77,10%

B. Occupational health and safety risks

Safety risks are those involving contact with machinery and infrastructure, as well as the processes and procedures involved, linked to them. These include risks of mechanical origin (contact with moving, cutting, pressure elements, etc.), risks of electrical origin, risks of ergonomic origin (postures, overexertion, among others) and all those related to processes and machinery and infrastructure.

— *Physical risk.* It is the risk caused by the presence of physical agents. These can be: noise, temperature, extreme pressures, radiation, among others. It is necessary for the responsible personnel to be familiar with these physical agents and understand their potential harmful effects. It should be clarified that the harmful effects of physical agents can be felt immediately or after long periods of time.

— *Chemical risk.* It is the risk presented by the use of chemical substances that have the potential to create serious health problems in the absence of proper use. These substances can be: dusts, fibers, metallic fumes, aerosols, chlorine gases, vapors, etc.

— *Biohazard.* Exposure to biological agents that may pose a threat to employees due to possible exposure to infectious agents. Agents that cause infections include bacteria, viruses and to a lesser degree fungi and parasites. These can be transmitted by inhalation, injection, ingestion or skin contact.

— *Fire and explosion.* Due to the seriousness of this hazard, it has been considered as a separate criterion from the other hazards mentioned. This hazard occurs when using substances that generate gases or vapors which, when in contact with combustible substances, may cause fire or explosion.

C. Corroboration of the diagnosis of the sanitation company according to OHSAS 18001.

It is necessary that the corroboration list of the diagnosis of the current state of the company SEDA-JULIACA complies with the international standard OHSAS 18001; for this purpose, the information gathered in the review of documents and interviews with the company's administrative and operational personnel was used.

The OHSAS 18001 standard specifies the requirements for an OSHMS, the design of which requires the resources available to the organization that wishes to be certified. This design is carried out by defining specific procedures and instructions in the company's scope of action.

Review of documentation

From the company's management, the existence of several documents was evidenced, among them:

— Organization and Functions Manual (MOF);
— Equipment manual;
— Safety manual;
— Operating procedures;
— Confidentiality and information security procedures;
— Work formats;
— Records (personnel files, corrective actions in non-conforming services).

It should be noted that the documents found do not follow the guidelines established by the OHSAS 18001 standard regarding document and record control (EPS ILO S.A., 2020).

Staff interviews

The results of the interview with the personnel involved in operations management were aimed at corroborating the company's current procedures and identifying those aspects of the regulations that are complied with.

Checklist results

From the collection of information, partial results were obtained for each requirement of the OHSAS 18001 standard and the total percentage of compliance.

Table 6. Scores established by the OHSAS 18001 standard

Requirements	Actual points	Total points	Compliance % Compliance % Compliance % Compliance % Compliance
4.1 General requirements	4	10	40
4.2 OSH Policy	5	11	45
4.3 Planning	3	19	16
4.4 Implementation and Operation	14	41	34
4.5 Verification	9	28	32
4.6 Management Review	3	7	43
Total	35	116	30%

Source: Spanish Association for Standardization and Certification (Ed.). *OHSAS 18001:2007. Occupational health and safety management systems - Requirements*, cit.

Table 6 shows the results of the level of compliance for each requirement of the OHSAS 18001:2007 standard; where the lowest aspect is in compliance with requirement *4.3: Planning*. The highest aspects are seen in compliance with requirements *4.2 Requirements OSH Policy* and *4.1 General requirements*.

Figure 16. OHSAS 18001 requirements met.

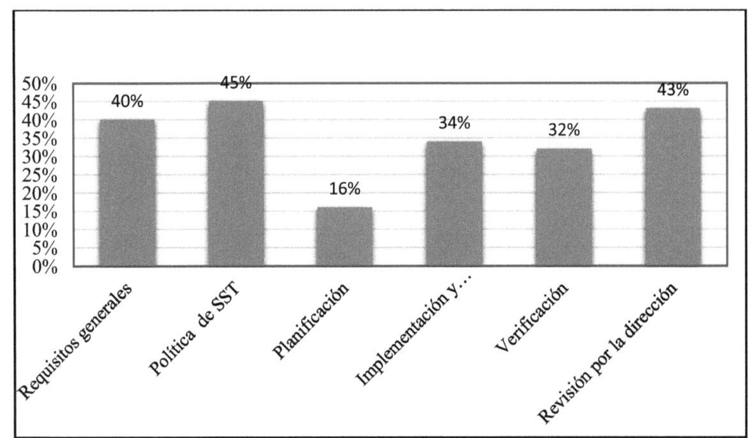

According to Figure 16, it can be seen that requirement *4.3 Planning* is the lowest with a 16% level of compliance, here the IPERC, legal requirements, objectives and programs are considered.

In addition to the general list, it was found that the level of compliance with the requirements for the standard applied in the sanitation company is 30% (see Figure 17).

Figure 17. Graphical representation of compliance

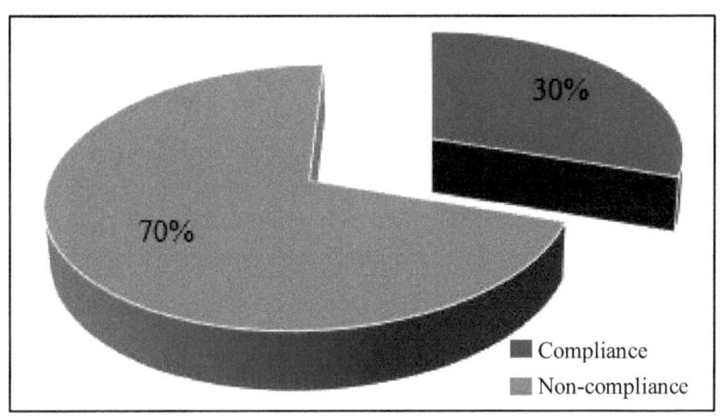

D. Implementation proposal criteria

Given the comparison of the OHSAS 18001 standard with D.S. 005-2012-TR, it can be seen that the "Basic" category includes compliance with Law No. 29783 and its regulation D.S. 005-2012-TR.

In this comparison it can also be seen that topics such as *General Provisions, Committee, Policy, Planning, OSH Development*, among others, are linked. This is the case with requirement 4.3 of the OHSAS 18001:2007 standard presented in Table 7.

Table 7. Requirements 4.3 of the OHSAS 18001 standard compared to the regulation D.S. 005-2012.

	OHSAS 18001:2007	Item number.	Regulation D.S. 005-2012
4.3	**Planning**	76 to 78	Planning and implementation SGSST
4.3.1	Planning for hazard identification, risk assessment and determination of controls	79 to 84 26 to 37	Planning and development Organization of the SGSST

4.3.2	Legal and other requirements	23 to 24 25	OSH Management System OSHMS Policy
4.3.3	Objectives and programs	79 to 84	Planning and development

Source: Spanish Association for Standardization and Certification (Ed.). *OHSAS 18001:2007. Occupational health and safety management systems - Requirements*, cit.

According to the OHSAS 18001 standard, the ESMS requirements must be implemented, which are broken down into elements 4.1, 4.2, 4.3, 4.4, 4.5 and 4.6.

Requirements 4.1 and 4.2 are related to general requirements and OSH policy (see Table 8).

Table 8. Requirements 4.1 and 4.2 of the OHSAS 18001 standard

	OHSAS 18001:2007	**Art.**	**Regulation DS-005-2012**
4	OSH Management System Requirements	38 to 73 74 to 75	The OSH committee or supervisor Rules of Procedure
4.1	General requirements		
4.2	OSH Policy	5 25	National OSH policy OSH Management System Policy

Source: Spanish Association for Standardization and Certification (Ed.). *OHSAS 18001:2007. Occupational health and safety management systems - Requirements*, cit.

Therefore, in the first phase of the proposal, the objectives must be established and in the second phase, which deals with the implementation of the OHSMS according to the OHSAS 18001 standard, the hazards are identified, risks are evaluated and control actions are specified.

In this sense, the policy, objectives, goals and management programs are determined, the control of documents and records is continued, the "SGSST Manual" is structured, and the implementation and documentation of procedures required by the standard that demonstrate the control of the system for each process executed in the company is carried out.

Preliminary actions of the company

Purpose

The company's purpose is focused on "providing quality drinking water and sewerage service".

2. Mission

To provide quality drinking water and sanitation services that protect the environment at the local level, with commercial efficiency, continuous communication, technological development and sustainability.

3. Vision

To become a leading company in the region providing quality assured services.

4. OSHMS and strategic objectives

Strategic planning is a very useful tool for companies today. In order to be sustained and successful, it requires a skill that can be derived from how well you know your company.

Among its benefits it is indicated that:

— It allows establishing the company's direction, objectives, priorities, goals and strategies.
— He has a rigorous knowledge of the current reality of the company and the environment that influences it.
— It frames quality improvement within a realistic, objective and feasible plan.

— It allows to involve and sensitize all the company's collaborators in the plans and objectives.

— Aligns activities and optimizes the use of resources in search of greater efficiency and achievement of objectives. An important part of establishing the opportunities and threats that exist in the organization is to carry out an internal diagnosis that allows to identify the strengths and weaknesses that exist and to see in a global way the whole environment.

It is an analysis that makes it possible to develop skills and competencies within any organization, as well as to make the best decisions based on the SWOT. This analysis should be carried out with the representatives of each area.

The functions established for occupational health and safety must be included in the company's strategic plan. In addition, the number of legal requirements involved increases over the years.

The interest of the organizations in demanding a management system to their suppliers; as well as the increase of the employees' expectations for a safe, healthy and pollution-free work environment. For this, it is necessary to establish as objectives the implementation of OSH requirements, provide a favorable work environment and the necessary training for the development of competent employees, provide sufficient tools and legal requirements to implement the OSHMS, improve the use of resources.

Requirements

Determine stakeholder needs

This refers to the incorporation of information (expectations and situations) from stakeholders in the implementation plan process.

Standards enforce compliance with government and regulatory requirements and expectations.

2. Identification of legal and regulatory resources

Its objective is focused on establishing the procedure for identifying, updating, accessing and monitoring compliance with regulations or other requirements related to OSHMS.

Thus, in order to proceed with such implementation, consideration should be given to codes of good corporate health and safety practice and public policy related to SEDA-JULIACA's functions.

3. Establishment of SGSST

Since SEDA-JULIACA carries out activities related to the treatment plant, OHSAS 18001, Law No. 29783 and Supreme Decree DS-005-2012-TR apply to these operational processes.

4. Definition of the implementation time

Senior management is responsible for communicating and ensuring that everyone is informed of the project start date.

It should be noted that, in the preparation of the Gantt chart on the project activities, in the case of the OSHMS, the activities for compliance should be contemplated according to the weaknesses highlighted during the initial diagnosis, in addition to the development and implementation of the plan, review and improvement of the OSHMS.

5. Processes within the scope of the OSHMS

Certain processes and operations are in place at the water treatment plant:

— Water catchment
— Cleaning of the river catchment area, sedimentation basins and filters
— Maintenance of electric pumps
— Treated water pumping by means of electric pumps
— Water quality control

This company must have a manual that includes everything that happens in the treatment plant, in order to provide data on the processes, indicators and formats of

OSH in your company. Likewise, the stipulations of D.S. 005-2012-TR will be taken into account to ensure that all necessary information is properly documented.

SEDA-JULIACA General Management Commitment

Concrete actions taken by the company's management demonstrate the company's commitment to safety and health in the work environment through specific activities, such as:

— Incorporate OSH-related topics into scheduled meetings.
— The work supervisor or CSST verifies occupational health and safety conditions.
— OSHMS validation is applied to ensure employee integrity.

The management review process should ensure that information is well managed and that it allows for evaluation. It is also required to include audits, statistical analysis of accidents, status of corrective and preventive actions, as well as results of previous management reviews. The management review analysis should be documented.

The review should integrate the need for changes in the system, including the policy and objectives (new or updated for continuous improvement), and establish the action plan to be followed.

The sanitation company SEDA-JULIACA, when submitting proposals and in harmony with the legal provisions, must specify the resources it will allocate for compliance with the SGSST.

Article 26 of Law No. 29783[58] states that the leadership of the OSHMS is the responsibility of the employer, who assumes the leadership and commitment of these activities in the organization, including: delegating the necessary functions and authority to the personnel in charge of the development, implementation and results of

---

[58] Congress of the Republic. *Law of Safety and Health at Work*, cf.

the OSHMS, who are accountable for their actions to the employer or competent authority.

This company also demonstrates knowledge of the OSH regulations in force in the country, which must be applied.

Thus, this company has defined a procedure to continuously identify and have access to applicable legal requirements. The organization keeps this information updated in the OSH committee.

E. Activities carried out in the company

This company strives to comply with the safety regulations applied to the different production areas, both in the wastewater treatment plant and in the networks, maintenance and offices (operations, commercial and engineering areas).

Treatment plant

It has unitary processes whose main purpose is to eliminate solids (particles and colloids). It is recommended that the companies carry out at least one control of turbidity, pH and aluminum, without exceeding the permissible limits prescribed by the corresponding entities. The processes carried out in each area are also detailed:

— *Catchment*. Water is collected by means of electrogenic pumps.
— *Sedimentation*. This is where coarse particles are retained by decantation.
— *Flocculation*. Hydraulic treatment unit for horizontal screens.
— *Filtration*. Sand filters are used.
— *Disinfection*. There is a chlorine dispenser.
— *Pumping*. It has three horizontal pumping units of 100 L/s each and one vertical pumping unit with a flow of 50 L/s.
— *Reservoirs*. It has six units: Cerro Colorado (2), Santa Cruz (3) and Independencia (1). Its total storage capacity is 10,745 m^3 .

The tasks performed within these areas require operating instructions for the correct handling of the machinery and maintenance of EPPS.

Figure 18. Cleaning of the catchment area on the banks of the Coata River in the Ayabacas sector.

This operation requires consideration of the following technical parameters:

1. Remove from the work area expendable elements that prevent the cleaning of banks.

2. Use protective equipment (helmet, gloves, among others).

3. Use rakes to remove elements from the bank.

Figure 19. Maintenance and cleaning of hydraulic cells and settling tanks.

The maintenance care of hydraulic cells and settling tanks is essential, therefore it is necessary:

1. Check cells and decanters for cracks.

2. Clean with scraper, brush and water jets.

3. Remove unnecessary objects and obstructions in the work space.

4. Use protective equipment.

5. The safe access platform is used to clean cells and settlers.

Figure 20. Power House

As observed in Figure 20, the work equipment is protected from the noise levels inside the generator equipment area, since they exceed the reference level. This operation requires consideration of the control and cleaning manuals.

Electric pumps

Regarding the handling and cleaning of electric pumps, it is important to perform certain procedures described below:

— Do not drive in excess of rated speed.
— Avoid contact with rotating equipment and install protection devices on electric pumps.
— Provide for proper processes to handle, install, operate and maintain the equipment.
— Proper lubrication of the pump is required for proper operation and the pump must be installed on a secure base to avoid vibration.
— If the noise level exceeds the standard level, wear hearing protection.

Floors (work surfaces) and corridors

The condition of the floors in the work environment is also relevant, so it is taken into account:

— Conditions of order, cleanliness and sanitation.

— Maintenance of pipes or drainage systems.

— No risk of falling.

— There are no protrusions such as rocks, concrete, among others, on the surface.

— Cracks must be plugged or armored.

Portable hand and power tools

Any type of tool or other element used are also of great importance, so it is necessary to have a proper storage, keep tools, electrical cables, ground connections and double insulation in good condition and operational, as well as train personnel to check the general condition of these equipment and maintain them for proper operation.

Warehouse

To keep this space functional, consideration should be given to:

— The order and cleanliness of the warehouse.

— Unobstructed access and transition areas.

— For loading and unloading both goods and raw materials, the vehicle must stop the engine and block the wheels by means of a block and the hand brake.

— Regularly inspect store shelves and replace defective products.

— Wear protective gloves and boots.

— Do not use objects weighing more than 25 kg for men and 15 kg for women. Always ask a colleague for help in these circumstances.

— When lifting an object, squat down, then bend your legs to lift the object, keep your back straight, grasp the object firmly and hold it as close to your body as possible. Do not turn at the hips or lift the load above the shoulders.

— Wear appropriate working clothes.

— When using a computer, sit with your back straight and your arms perpendicular to the table.

— Position the monitor so that the top of the screen is aligned approximately at eye level and avoid reflections from light and windows.

Circulation areas

For the maintenance of this area it is necessary to follow certain guidelines:

— Clean floors and post signs warning of slipping hazards.

— Aisles shall be kept neat and free of obstructions at all times.

— Do not use both hands when ascending or descending stairs. It is important that one hand is free to hold on to the handrail.

— If several products are to be moved, they must be visible from the top and from both sides when supported by the arms.

— Install emergency exit signs.

— Signal glass doors with a reflective strip, which shall be placed over the entire width of the door at 1.40 m from the floor.

Lighting

It is essential that several points be considered:

— Transit and work areas should be properly illuminated.

— Corridors, stairways and offices need lighting at all times.

— Prevent eyestrain.

— Emergency lighting should be installed on all routes.

General furniture

All furniture must meet certain parameters:

— Implement comfortable, well-fitting, height-adjustable and, if possible, stable and swivel chairs.
— Have a comfortable desk and durable plastic tablecloths.
— Do not place boxes, papers or other heavy items on top of cabinets, tables or filing shelves, as they may fall on any individual.
— Close furniture drawers.
— Heavy appliances or shelves are placed close to the walls.

Electrical records

Regarding this aspect, it is established that:

— Electrical repairs are only performed by authorized and trained personnel.
— Inspect electrical connections in the office for worn wires and plugs or cables that restrict employee movement.
— Prevent cables from being placed in employee circulation areas.
— When disconnecting the plug, be sure to hold the plug instead of pulling the cable.

Signage and labels

Arbitrary removal of warning signs or labels is prohibited and may result in fines and possible reprimand, suspension or dismissal, depending on the severity of the infraction.

Tasks involving high risks such as heights, agglomerations, high voltage machinery, gas intoxication and environmental contamination must be marked with warning and danger signs.

Similarly, emergency evacuation signs and safety zones are required in all spaces. Appropriate signage requires the use of established national color and symbol codes (INACAL, 2016).

Therefore, this company must have:

— Hazard warning signs in areas of imminent danger.
— All defective equipment should be marked with a label.
— The color-coded signs indicate certain conditions: imminent danger (red color), in modification (orange color), changing area (yellow color), safety and first aid (green color), general information (blue color).
— Label those containers and transports of harmful substances.

Lighting

If natural light is insufficient, artificial lighting will be installed in the work areas and the rest of the company's spaces. Artificial light is uniform, of a continuous intensity and sufficient to perform the functions with tranquility: while natural light enters through skylights, windows or walls made of light-transmitting materials to ensure homogeneous illumination.

Chemicals and fuels

It is important that all company personnel have appropriate protection to perform their duties, especially if working with toxic gases (such as chlorine gas), exposures to chemical compounds. On the other hand, the tank must be equipped with a pressure safety valve, with support and protect it from corrosion. It is also essential to control the temperature of the materials in order to avoid boiling.

Figure 21. Hazardous chemical agents

Source: EPS SEDA-JULIACA. *General company data*, cit.

Figure 22. Map of risk signaling in the company SEDA-JULIACA.

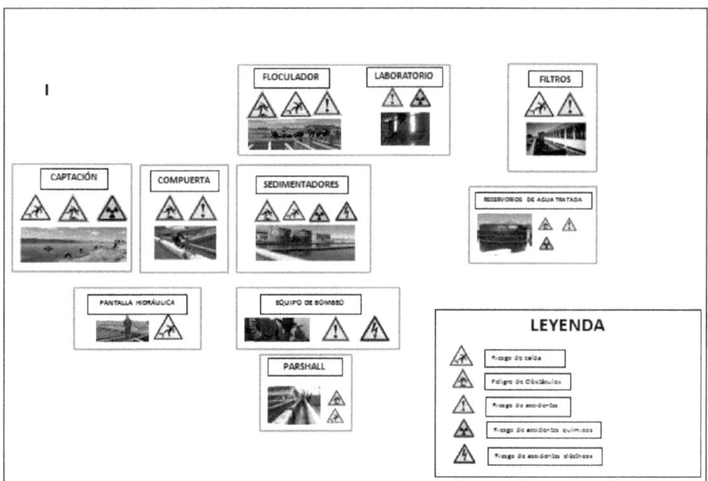

Source: EPS SEDA-JULIACA. *General company data*, cit.

Specific risks

This company takes into account that proper hazard management supports operational plans. In this regard, certain impact measures to be considered for flooding are noted:

— The Coata riverbed is eroded.

— Overflow generated from the water intake point to the treatment plant.

— Development of water intake networks (collection grids) for wastewater treatment facilities.

— The impact is generated from January through March.

In the event of a drought situation, the impact is as follows:

— The flow of the Coata River is reduced.

— Decrease in personnel due to reduced production.

— Since some machinery operates intermittently, sediment accumulates inside the pipes and, consequently, water quality deteriorates.

Electrical hazards

The materials used in all electrical equipment will be chosen according to the voltage, load and other conditions, provided they comply with those established in Peruvian electrical regulations (MINEM, 2006).

Each cable must be adequately insulated and securely fastened to the raceway with an escape route at each end in the event that tracks are located overhead or subway.

Distribution panels with exposed conductors and alternating currents or ground voltages exceeding 50 volts must be protected with appropriate barriers accessible only to authorized employees.

Similarly, when repairing a machine, safety precautions should be taken, such as removing fuses and using *lock out* systems to prevent other individuals from activating the machinery while it is in use.

Employees who repair electrical equipment require training to work with voltage.

In addition, leather gloves and insulated shoes without metal parts are required. Also, separate work stations, such as platforms or divided floors, are used.

Finally, all employees performing their duties as electricians must be trained in cardiopulmonary resuscitation or first aid if an electric *shock* is generated.

Risk of falling

The dynamics of the fall due to gravity, and the "falling" object shows a favorable acceleration and increases its velocity until it stops due to an external force. In short, these dynamics occur during the phases of free fall, acceleration and arrest.

Therefore, systems that prevent falls include: bucket-type lifts to modify street lighting and scissor lifts, pulley platforms, lifting systems with different types of winches (electric, pneumatic and manual) and with electric ladder rails.

Chemical hazards

According to Del Pezo[59] , human exposure to toxic gases (chlorine gas, poison gas) is considered an imminent hazard. If a worker inhales a concentration of this substance that exceeds 1000 ppm, it can cause death. It also has another type of impact on the employee's lungs, causing conditions such as chronic bronchitis.

Similarly, if the worker uses aluminum sulfate, he/she may develop various symptoms, such as:

— Irritation of eyes, gastrointestinal tract and skin from short-term exposure to this chemical.
— Affection of the central nervous system due to long-term exposure to this chemical compound.

Incident and accident investigation and analysis

---

[59] Otto Gabriel Del Pezo De la Cruz. "Modelo de gestión de seguridad y salud ocupacional para la empresa de agua potable, aguas de Península - Aguapen S. A." (Master's thesis). Ecuador, Universidad Politécnica Salesiana, 2013, available at [https://dspace.ups.edu.ec/handle/123456789/4829].

Whenever an occupational incident or accident occurs, most of the problems originate from a variety of reasons; it is rare that they result from one and the same reason. Therefore, it is important to identify the root cause of the problem to prevent the incident from reoccurring[60] .

So, from every investigation we obtain the details of what happened, the recognition of the origin of an accident, identification of risks, approach and execution, control system and demonstration of interest.

1. Inventory of jobs

First, a list of essential activities is created for all personnel involved in a specific task. In this case, there are operators in plant, pumping equipment, pumping equipment maintenance, hydraulic cell cleaning, water and sewer installation, and network maintenance.

Then, each function is divided into activities, so that it is possible to examine each of these and indicate its degree of importance or irrelevance. This is done by the supervisors as a group. The primary tasks of the operators in the treatment plant are:

— Execute water collection and drainage pumps and their respective maintenance.
— To carry out the operation of the hydraulic cells and settling tanks.
— Proceed with the care and cleaning of conventional filters.
— Clean each workspace.
— Execute the cleaning functions in the sewage networks.

2. Identification of critical jobs

Activities with a history of loss, through personal injury, property damage or diminished product quality, are categorized according to their criticality.

---

[60] CONICYT. *Standards manual: biosafety and associated risks*. Chile, Fondecyt-CONICYT, 2018.

Because this program is predictive rather than reactive, it is essential to integrate possible losses into the task, even if historical data exists. For this, several questions need to be asked:

— Can this task, if not executed correctly, result in a serious loss while it is being performed?

— Can this task, if not executed correctly, result in a serious loss after it has been performed?

— How severe can the loss be (How serious can the injuries be)?

— How often is this expected to occur?

The regularity of appearance is specified by several factors, the most important of which are:

— The number of times a task is executed during a period of time (repetition).

— It is possible that a loss will be generated after executing the activity (probable loss).

It is essential to note that there are varying degrees of importance and that any task worth performing is, in fact, important to some degree. In this sense, a criticality scale can be developed for the system.

3. Work at heights

The use of seat belts is mandatory when working at an altitude of 1.80 m, due to the risk of free fall or other disadvantages mentioned above.

Likewise, employees working at heights must receive special training in the use of safety belts and harnesses to prevent improper installations.

4. Emergency response

All those facilities, operations and activities developed in the company SEDA-JULIACA, within its environment are technically qualified as "moderate risk" and, despite the occurrence of an emergency situation, they cannot be controlled by the company's own employees and require internal resources. Therefore, their qualities and impacts must be managed within an organizational framework that facilitates an efficient response.

5. Contingency plan

It is of utmost importance that this plan be implemented in each area to prepare for and minimize damages that may occur during an emergency.

This plan requires a person in charge of its execution, evaluation and dissemination, as well as the training of the different emergency brigades.

To ensure the training of employees, exercises are organized taking into account the non-interruption of operations at the treatment plant.

This plan also provides added value in terms of safety in case of emergency and must be developed with sufficient standards to be respected and to allow the evaluation of its results in a flexible and efficient manner. Similarly, these procedures should be taken into account:

— Plan of evacuation routes in emergency situations.
— Worker rescues and medical supervisions are carried out.
— List of staff emergency contact names, including address and phone numbers.

Emergency situations that require coverage include explosions, fires, floods, earthquakes, personal injuries, etc.

It is important to clarify that telephone numbers should be published without restrictions in the main security rooms, offices, areas where a large number of employees are located.

6. First aid kit

First aid kits are provided with medicines and other basic necessities, the type and availability of which will be specified on a case-by-case basis. Similarly, emergency personnel will receive training to ensure a rapid and effective response.

Among the recurring emergency situations in the companies analyzed are floods, fires, earthquakes (which may require the evacuation of personnel in an appropriate and effective manner).

This emergency plan is effective because it ensures that personnel are familiar with the facilities and proper evacuation procedures. Tools used to communicate these procedures include evacuation plans and signage.

7. Non-conformities, incidents, accidents and occupational diseases

Accidents, incidents and occupational diseases must be reported to the person responsible for the human resources area, including the injured person, a close colleague or immediate supervisor.

Emergency services are responsible for transporting people to the nearest hospital to receive the required treatment.

Accidents must be reported to the OSH chairperson and immediate supervisor through the HR assistant within 24 hours. It should be understood that this HR assistant is also responsible for documenting reports, investigating causes, implementing corrective and preventive actions.

While the immediate supervisor is responsible for investigating the accident, determining the causes and taking corrective or preventive measures within the first five days.

Meanwhile, the OSH chairman is in charge of training the related parties regarding reporting, investigation of the reason and implementation of countermeasures. The committee, together with OSH officers, are in charge of inquiring about the accident involving death or disability. In addition, accidents with such tragic results are reported to the Ministry of Labor at the respective headquarters.

Internal audits

The company must apply an implementation method to carry out the internal OSH assessment, and thus establish whether it complies with the standards and requirements of a management system determined in that company or provided by legal regulations.

According to OHSAS 18001, organizations should plan internal audits to monitor compliance and evaluate the effectiveness of their OHSMS. It should be added that the assessment should have a schedule based on the importance and context of the process to be audited. Deficiencies identified during an audit should be corrected in a timely manner. These standards require an effective audit process, performed by competent personnel, using a defined procedure (see Figure 23).

Figure 23. Continuous improvement

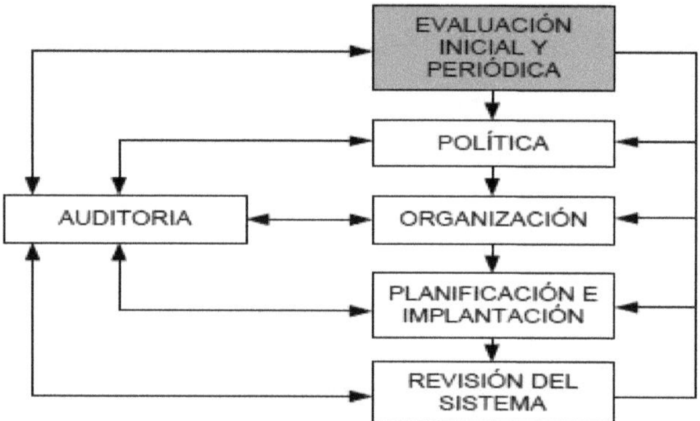

Source: Spanish Association for Standardization and Certification (Ed.). *OHSAS 18001:2007. Occupational health and safety management systems - Requirements*, cit.

Figure 24. The audit process

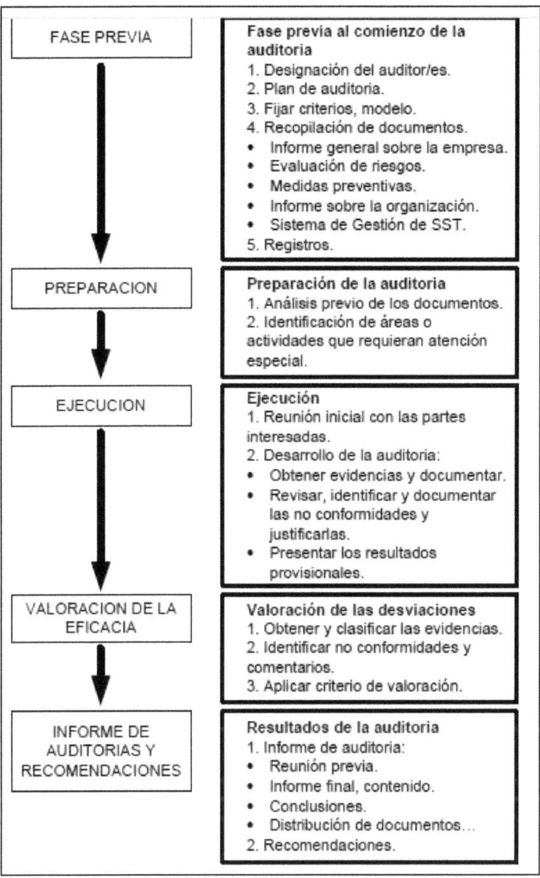

Source: Spanish Association for Standardization and Certification (Ed.). *OHSAS 18001:2007. Occupational health and safety management systems - Requirements*, cit.

XI. Analysis and interpretation of results

A. Initial diagnosis

The physical, chemical, biological, ergonomic and psychosocial risks of the operations were analyzed.

The results show the level of compliance by clauses of the OHSAS 18001 standard in the SEDA-JULIACA company.

At the end of the application of the data collection instruments, a score was obtained for each requirement of the OHSAS 18001 Standard; the total is 30% compliance and 70% non-compliance, therefore, measures must be taken in the OSHMS proposal.

In this sense, the requirements requested in the OHSAS 18001 standard are presented, according to their non-conformities.

General requirements

The evidence is as follows:

— Continuous improvement is not practiced.
— Empathy between employees and employers is not promoted.
— Lack of mechanisms to strengthen feedback between employees and employers.
— There are no mechanisms for recognizing personnel who are proactive in continuous improvement.
— Methodologies for continuous improvement are not used.

OSH Policy

The evidence is as follows:

— There is no documented OSH policy.
— Decisions are not made on the basis of audits.
— The employer does not take the lead on OSH.
— The employer is not committed to OSH management.
— There is no adequate budget.
— The requirements for each position in terms of OSH have not been defined.

Table 9. Company principles

Guidelines	Indicator			Compliance
I. Commitment	Involvement	Yes	No	Remarks
	The employer is committed to occupational safety and health.	X		
Principles	Consistency is achieved between what is planned and what is carried out		X	
	Continuous improvement is practiced		X	
	Improves self-esteem and fosters team work	X		
	Promoting a proactive risk prevention culture	X		
	Empathy from the employer to the employee or vice versa is encouraged.		X	
	There are mechanisms in place to recognize proactive personnel for continuous improvement.		X	

	An assessment of the primary risks that generate the most losses is maintained	X		
	A methodology for continuous improvement is used.		X	
	The participation of trade unions, whose representatives take part in decisions on occupational safety and health, is encouraged.		X	

Planning

The evidences are:

— There is no initial OSH assessment.
— There is no OSH planning.
— There are no procedures to identify hazards and evaluate risks.
— There are no OSH objectives.
— There are no preventive measures.

Implementation and Operation

It is evident that:

— The employer does not anticipate that exposure to physical agents will harm the worker.
— The employer does not transmit information to workers regarding OSH.
— The OSH program is not reviewed.
— OSH courses are not documented.

— There are no plans and procedures for dealing with incidents and emergency situations.

— Incidents and accidents are not reported.

Verification

The following is evident:

— There is no OSH monitoring.

— Medical examinations are not taken into account for corrective actions.

— No investigations are conducted on occupational accidents.

— Corrective and preventive measures are not taken.

— Operations and activities that are associated with a risk have not been identified.

— There are no control procedures.

Management Review

This evidence is shown:

— Management does not periodically review and analyze the OSHMS.

— There is no evidence of review or audits for management to achieve the intended purposes.

— There is no evidence of a review of the OSHMS performance (see table 10).

Table 10. Management review

VII. Management review		Yes	No
	Senior management:	X	

	Periodically reviews and analyzes the management system to ensure that it is appropriate and effective.			
Measureme nt	Arrangements made by management for the continual improvement of the occupational health and safety management system should take into account: • The company's occupational health and safety objectives. • The results of hazard identification and risk assessment. • The results of hazard identification and risk assessment. • The results of monitoring and measuring efficiency. • Investigation of work-related accidents, illnesses and incidents. • The results and recommendations of audits and evaluations performed by the company's management. • The recommendations of the safety and health committee, or of the safety and health supervisor, or of the safety supervisor • Changes in legal regulations. • The results of the health protection and promotion program.	X		

	The continuous improvement methodology considers: • Identification of deviations from accepted safe practices and conditions. • The establishment of safety standards. • Periodic measurement and evaluation of performance against standards. • Correction and recognition of performance.			
Continuous improvement management	The investigation and audits allow the company's management to achieve the intended purposes and to determine, if necessary, changes in the policy and objectives of the management system.		X	
	The investigation of accidents, illnesses and incidents should identify: • Immediate causes (substandard acts and conditions). • Immediate causes (personal factors and work factors). • Deficiency of the safety and health management system, for the planning of the relevant corrective action.	X		
	The employer has modified the occupational risk prevention measures when they are inadequate and insufficient to ensure the safety and health of workers.		X	

B. Aspects established for the company's health and safety management system.

Among the aspects that should be included in an OSH management system are: carrying out the IPERC; making an annual OSH plan; planning meetings, inspections, cleanings, trainings and audits; providing maintenance to machinery and audits; as well as the internal safety and occupational health regulations.

1. Hazard identification and risk assessment (IPERC)

In order to identify risks, the company must specify procedures for recognizing ongoing hazards, assessing risks and determining the necessary control measures (see Table 11).

It should include:

— Routine and non-routine activities
— Activities of personnel accessing the treatment plant facilities.

Table legend 11

Exp.   = Number of exposures

TD Exp = Exposure time

F   = source

M   = medium

I   = Individual

C   = consequences

E   = Exposure

GP   = degree of danger

INT. 1 = GP Interpretation

FP   = weighting factors

GR   = degree of impact

INT. 2 = GP interpolation

P   = Probability

Table 11. Diagnosis of working conditions at the treatment plant.

Process	Operation	Risk factor	Source of risk	Possible effects	# from Exp.	Exposure time (hours)	Current control			valuation								Remarks
							f	m	i	C	P	E	GP	INT 1	FP	GR	INT 2	
Treatment of **water**	Coagulation	Physicist	Treatment Module	Drop-off	3	1			x	10	1	6	60	under	1	60	under	Lack of handrails
	Sedimentation	Physicist	decanters	Drop-off	5	2			x	10	1	6	60	under	1	60	under	No railings
		Chemist	decanters	Suspended gases	4	4			x	1	7	6	42	under	1	42	under	Gas emissions
	Filtration	Physicist	Filtration wells	Drop-off	3	1		x		10	1	6	60	under	1	60	under	No handrails
	Chlorination	Chemist	Dosing room	Gas inhalation	4	1		x		6	7	6	252	under	1	252	under	Chlorine odor
	Storage	Physicist	reservoirs	Drop-off	1				X	1	7	6	42	under	1	42	under	No railings

Source: EPS SEDA-JULIACA. *General company data*, cit.

Table 12. Hazard identification, risk assessment matrix (IPERC)

AREA	TASK O ACTIVITY	DANGER	RISK	TYPE OF ACTIVITY	CONTROL EXISTING	RATE OF EXPOSED PERSONS (IE)	INDEX OF RISK EXPOSURE (IF)	TRAINING RATE (ICE)	PROCEDURAL INDEX EXISTING (IPT)	PROBABILITY INDEX IP=IE+IF+IC+IPT+IPT	SEVERITY INDEX (IS)	MAGNITUDE OF OCCUPATIONAL RISK IML=IPXIS	CONTROL MEASURES TO BE IMPLEMENTED
	Water catchment of the Coata River	River impoundment during rainy season	Falling into the river	Non-routine	Signage	4	2	2	3	11	2	22	Signage, report river flooding, do not approach the river bank.
	It is pumped to the treatment plant by means of electric pumps.	Equipment without protective guard	Electric shock, trapping			3	2	2	3	10	2	20	Warning signs for moving equipment and placement of protective guards.

89

FLOOR	Maintenance and cleaning of hydraulic screen	Falling to the hydraulic screen	Hits	Routine	Signage	2	2	2	2	8	2	16		Warning signage and use of helmet, boots, waterproof clothing
	Cleaning of settling tanks	Fall or slip	Hits	Non-routine	signage	3	2	2	2	9	2	18		Warning signs for the use of helmets and boots
	Changing the chlorine gas tank to the dosing unit	Gas leakage	Gas inhalation	Non-routine	signage	2	2	2	2	8	3	24		Warning signs for the use of masks and securing the ball during the changeover.
Distribu tion network s	Breaks and leaks in water mains and potable water distribution networks	Leaks Aniego Flooding on public roads	Automob ile accident	Non-routine	Signage	2	1	2	3	8	3	24		Signage, water leakage control programs, water quality control programs
	Cutting or mechanical	Unsuitabl e	Injuries blows	Routine	Signage	2	1	2	3	8	3	24		Put up signs, gates and tapes

	breakage of pavement	equipmen t											with no trespassing notices, use of overalls, helmet, boots and gloves.
Distribu tion network s	Leaks	Sidewalk sinking	Falls	Routine	Signage	3	2	3	2	10	2	20	Place signs in the work area and use helmet, overalls, gloves and boots.

Source: Ministry of Labor and Employment Promotion. *Approval of reference formats that contemplate the minimum information to be contained in the mandatory records of the Occupational Health and Safety Management System.* Ministerial Resolution No. 050-2013-TR of 14-03-2013. Lima, Peru, MTPE, 2013, available at [https://www.gob.pe/institucion/mtpe/normas-legales/288031-050-2013-tr].

2. Annual OSH program for the sanitation company.

The program determines the deadlines and goals to be achieved as part of the objective of the planning proposal, among which are: drills, inspections, training, equipment maintenance and induction of new workers.

3. Training Program

Based on OSH regulations, the risk assessment has been taken into account and an annual OSH training program has been created, adapted to the requirements of each activity carried out in the company, the job description and technical needs, including external courses. The Safety Management is responsible for its preparation, compliance and evaluation, in coordination with the heads of the different areas, taking into account the defined guidelines. Standard SSMAT-P03.03 "Training" considers the following aspects:

- detection of training needs;
- scheduling of training topics;
- awareness
- and training within the workday;

Keep in mind that training records are kept by the person responsible for training and each area.

Table 13. Activities related to critical risks

ITEM	CRITICAL RISKS	OPERATIONS / RELATED ACTIVITIES
1	Electrocution	Main electrical installation and power house
2	Hit-and-run	Transportation of material and leveling of floors with tractor, maintenance of trucks, transportation of personnel, transportation of fuel.
3	Fall of persons	Construction and repair of decanter cracks, maintenance and repair of equipment and electrical installations. Excavation activities.
4	Gassing	Maintenance and cleaning activities for chlorine and fuel tanks.
5	Sight injuries	Maintenance process for generator equipment or pumps.
6	Low back pain	Process of maintenance and repair of equipment. Manual transfer of materials in warehouses.
7	Crushing	Maintenance process for pumps or vehicles
8	Cut	Cuts due to mishandling of objects with sharp or cutting edges, in maintenance processes, welding, civil works, transfer of materials.
9	Noise-induced hearing damage	Operation of compressors, power house generators, drilling, etc.

10	Respiratory diseases due to dust exposure	Transportation of materials, drilling, earthworks, access roads:

Source: Ministry of Labor and Employment Promotion. *Approval of reference formats that contemplate the minimum information to be contained in the mandatory records of the Occupational Health and Safety Management System*, cit.

Table 14. Inspections

TYPE	FREQUENCY	LOCATION	PERFORMED BY	RESPONSIBLE FOR COMPLIANCE WITH RECOMMENDATIONS	REPORTING PROCEDURE	FOLLOW UP
CONTINUA	Daily	High Risk Areas	Capture	Area managers	To area managers and	Take immediate action or initiate report for corrective action
		Workshops	Maintenance			
	Weekly					Take immediate action or initiate report for corrective action
		Wells, decanters	Maintenance	Supervisors	Supervisors	
		Power House	Warehouse			Record observations in the Powder Dust Dust Annex.
	Monthly	Equipment	Maintenance	Supervisors	Supervisors	The following must be reported within 10 days
			Security			

		Order and Cleanliness	All areas			Safety area of the corrective actions taken and their results.
SCHEDULED	**Quarterly**	Unit	Sub-management General Supervisor Security	Area Supervisor	Report observations Made to the supervisor General	Draw up a program for the removal of the observations, indicating: responsible parties, measures to be taken and deadlines.
MONTHLY	**Undetermined**	Unit	Security Committee	Area Superintendents Supervisors	Report observations Made to the superintendence General	Draw up a program for the removal of the observations, indicating: responsible parties, measures to be taken and deadlines.

Source: Ministry of Labor and Employment Promotion. *Approval of reference formats that contemplate the minimum information to be contained in the mandatory records of the Occupational Health and Safety Management System*, cit.

Table 15. Simulation program

TREATMENT PLANT																
N.°	DESCRIPTION	RESPONSIBLE	STATUS	Jan-18	Feb-18	Mar-18	Apr-18	May-18	Jun-18	Jul-18	Aug-18	Set-18	Oct-18	Nov-18	Dec-18	TOTAL
1	FIRE DRILL	Chief SSO	SCHEDULED	1				1					1			3
2	EARTHQUAKE SIMULATION	Chief SSO	SCHEDULED						1					1		2
3	MOCK ACCIDENT DRILL	Chief SSO	SCHEDULED				1				1				1	3
4	SPILLAGE OF HAZARDOUS SUBSTANCES	Chief SSO	SCHEDULED						1							1
TOTAL NUMBER OF DRILLS			SCHEDULED	1	0	0	1	1	2	0	1	0	1	1	1	9
			EXECUTED	0	0	0	0	0	0	0	0	0	0	0	0	0
			PENDING	1	0	0	1	1	2	0	1	0	1	1	1	9

Source: Ministry of Labor and Employment Promotion. *Approval of reference formats that contemplate the minimum information to be contained in the mandatory records of the Occupational Health and Safety Management System*, cit.

Table 16. Infrastructure program

INFRASTRUCTURE PROGRAM																
TREATMENT PLANT																
N.°	INSTALLATION	RESPONSIBLE	STATUS	Jan-18	Feb-18	Mar-18	Apr-18	May-18	Jun-18	Jul-18	Aug-18	Set-18	Oct-18	Nov-18	Dec-18	TOTAL
1	OFFICES	Area Manager	SCHEDULED	1		1		1		1		1		1		6
			EXECUTED													
2	GENERAL STORE	Area Manager	SCHEDULED	1	1	1	1	1	1	1	1	1	1	1	1	12
			EXECUTED													
3	POWERHOUSE	Area Manager	SCHEDULED	1	1	1	1	1	1	1	1	1	1	1	1	12
			EXECUTED													
4	FIRE EXTINGUISHER INSPECTION	Chief SSOMA	SCHEDULED	1	1	1	1	1	1	1	1	1	1	1	1	12
			EXECUTED													
5	EMERGENCY TEAMS	Paramedics	SCHEDULED	1		1		1		1		1		1		6
			EXECUTED													
TOTAL INSPECTIONS			SCHEDULED	5	3	5	3	5	3	5	3	5	3	5	3	48

| | EXECUTED | | | | | | | | | | | | |
| | PENDING | 5 | 3 | 5 | 3 | 5 | 3 | 5 | 3 | 5 | 3 | 5 | 3 | 48 |

Source: Ministry of Labor and Employment Promotion. *Approval of reference formats that contemplate the minimum information to be contained in the mandatory records of the Occupational Health and Safety Management System*, cit.

Table 17. Equipment and tooling inspection program

EQUIPMENT AND TOOLS INSPECTION PROGRAM																
TREATMENT PLANT																
N.°	UNITS AND EQUIPMENT	RESPONSIBLE	STATUS	Jan-18	Feb-18	Mar-18	Apr-18	May-18	Jun-18	Jul-18	Aug-18	Set-18	Oct-18	Nov-18	Dec-18	TOTAL
1	ELECTRO PUMPS	Chief SSOMA	SCHEDULED	1	1	1	1	1	1	1	1	1	1	1	1	12
			EXECUTED													
2	VEHICLES AND MOBILE EQUIPMENT	Chief SSOMA	SCHEDULED	2	2	2	2	2	2	2	2	2	2	2	2	24
			EXECUTED													
3	ACCESSORY ELEMENTS	Chief SSOMA	SCHEDULED	1	1	1	1	1	1	1	1	1	1	1	1	12
			EXECUTED													
4	HAND AND POWER TOOLS	Chief SSOMA	SCHEDULED	1	1	1	1	1	1	1	1	1	1	1	1	12
			EXECUTED													
	TOTAL INSPECTIONS		SCHEDULED	5	5	5	5	5	5	5	5	5	5	5	5	60

EXECUTED													
PENDING	5	5	5	5	5	5	5	5	5	5	5	5	60

Table 18. External trainings

Occupational Health and Safety Policy	To prevent and control occupational hazards and occupational health risks to the integrity of our employees and to minimize the environmental impacts on our processes and populations in the work environment.															
General Objective	**Occupational Health and Safety at Work System Planning Proposal**															
Specific objective	To achieve that 90% of the staff attends the scheduled external trainings and obtains a satisfactory rating.															
Goal	% > 90															
Indicator	Efficiency (training coverage): total number of people trained vs. planned Effectiveness (quality of training): actual score obtained / score expected															
Budget	6000 nuevos soles															
Resources	Commitment of all Area Managers															
NO.	Description of the activity	Responsible of execution	Area	YEAR: 2018											Date of verification	Status (completed, pending, in process)	Remarks
				E	F	M	A	M	J	J	A	S	O	N	D			

No.	Activity	Responsible	Area														Frequency	Status	Observations
01	Send in advance the Attendance Programs of the personnel to external trainings.	Area managers	All areas	X	X	X	X	X	X	X	X	X	X	X	X	Monthly	In process	None
02	Follow up on external training programs sent by area managers.	Safety Management and the OSH Joint Committee	All areas	X	X	X	X	X	X	X	X	X	X	X	X	Diary	In process	None
03	Conduct surveys on the quality of the exhibitor and the subject matter presented.	Security Management	All areas	X	X	X	X	X	X	X	X	X	X	X	X	Diary	Pending	None
04	Analyze at the end of each course the fulfillment of the efficiency and	Joint Safety Committee and Area Managers	All areas	X	X	X	X	X	X	X	X	X	X	X	X	Monthly	In process	None

	effectiveness of its applicability.																			

Source: Ministry of Labor and Employment Promotion. *Approval of reference formats that contemplate the minimum information to be contained in the mandatory records of the Occupational Health and Safety Management System*, cit.

Table 19. Ergonomic risk control

Occupational Health and Safety Policy	Preventing and controlling occupational hazards and occupational health																
General Objective	**Proposal for the Planning of an Occupational Safety and Health Management System in the Workplace**																
Specific objective	Controlling ergonomic risks																
Goal	Control and prevention of ergonomic risks																
Indicator	Comfort in the work area																
Budget	1000 nuevos soles																
Resources	Commitment of all area managers to control ergonomic risks.																
NO.	Description of the Activity	Responsible of Execution	Area	YEAR 2018												Date of verificati on	Status (Completed , Pending, in process)	Remarks
				E	F	M	A	M	J	J	A	S	O	N	D			

01	Preventive maintenance of equipment and tools in the workstations, according to use and time.	Security Management, and Area Managers	All areas	X	X	X	X	X	X	X	X	X	X	X	X		Monthly	In process	None
02	Identification of factors, evaluation of work areas, position on site, recommended limit load.	Security Management Area Managers	All areas	X	X	X	X	X	X	X	X	X	X	X	X		Monthly	In process	None
03	Training and sensitization of all personnel on postural positioning in	'	All areas	X	X	X	X	X	X	X	X	X	X	X	X		Monthly	In process	None

the workplace, repetitive movements, work-rest, perceptual and mental overload.	Area Managers / Security Management																

Source: Ministry of Labor and Employment Promotion. *Approval of reference formats that contemplate the minimum information to be contained in the mandatory records of the Occupational Health and Safety Management System*, cit.

Table 20. Induction, training and emergency drills record format.
emergency

REGISTRY NO:	RECORD OF INDUCTION, TRAINING, EDUCATION, TRAINING AND EMERGENCY DRILLS.				
EMPLOYER'S DATA					
COMPANY NAME OR CORPORATE NAME	RUC	ADDRESS (Address, district, department, province)		TYPE OF ACTIVITY	NUMBER OF WORKERS IN THE WORK CENTER
EPS. SEDA - JULIACA		TREATMENT PLANT			
MARK X					
INDUCTION		TRAINING		TRAINING	EMERGENCY DRILL
TOPIC 9	EARTHQUAKE SIMULATION				
DATE					
NAME OF TRAINER OR TRAINER	SAFETY ING. SECURITY ING.				
NO. HOURS					
SURNAMES AND NAMES OF THOSE TRAINED	DNI NO.	AREA		SIGNATURE	REMARKS
1.					
2.					
3.					
4.					
5.					
6.					
7.					
8.					
9.					
10.					
11.					

12.				
13.				
14.				
15.				
16.				
17.				

Source: Ministry of Labor and Employment Promotion. *Approval of reference formats that contemplate the minimum information to be contained in the mandatory records of the Occupational Health and Safety Management System*, cit.

Table 21. Written Procedure for Safe Work (PETS)

TREATMENT PLANT SEDA-JULIACA	WRITTEN SAFE WORK PROCEDURE	CODE	PETS PG 03.01
		VERSION	
	TRAINING		
		DATE	20/01/2018

The model template is presented with the procedure for each section:

1. Objective

Describe and detail the work methodology at the treatment plant.

2. Scope

The procedure begins with the entry of new personnel to the sanitation company SEDA-JULIACA.

3. Legal Basis

3.1. Law No. 29783

3.2. OHSAS 18001:2007 Standard

3.3. SEDA-JULIACA Company Standards

4. Definitions

4.1. Training: activity that consists of transmitting theoretical and practical knowledge for the development of competencies, capabilities and skills regarding the work process, risk prevention, safety and health.

4.2. Training plan: a tool to record the training needs of personnel in line with the needs of the job.

4.3. External training: training provided by external personnel. It can be conducted inside or outside the organization's facilities.

4.4. Internal training: training conducted by the organization's permanent staff.

4.5. Induction: This is mandatory training, preferably for new personnel joining the organization. The topics covered are: occupational health and safety, social responsibility and the environment.

4.6. Induction of visitors: Induction is given to all visitors prior to entering the organization.

5. Responsibilities

Human Resources Manager

- — Ensure compliance with this procedure.
- — Follow-up of the process in its different stages.

5.2. Area Manager.

- — Identify the training needs of workers.
- — Coordinate with the management area the scheduling and execution of training according to the priority or need of the work area.
- — Ensure the attendance of new personnel under their charge to the induction participation.

— Ensure that area managers attend operational risk courses.

5.3. Supervisor

Provides specific induction to new workers under his responsibility and ensures the attendance of new personnel under his responsibility to the participation of the specific module for operational risks.

5.4. Worker

Attends training scheduled by their supervisor on the dates they are instructed.

6. Development

Table 22. Activities related to training

Activity	Responsible	Description	Registra tion
Training	Coordinators Management	1. The attendance of workers and exhibitors will be duly controlled with the participation register (SSOMA FG.01) or by the reports issued. 2. Training plans can be individual or group-based and can be conducted on or off-site. It is important to contribute to compliance with the occupational health and safety policy, as well as with the instructive procedures and requirements of the management system.	

		3. Awareness and sensitization of workers will be done through five-minute talks, group meetings and induction.	
		4. Group meetings are conducted in accordance with the group meeting procedure.	
		5. Conduct evaluations that measure the level of knowledge acquired from the training and induction courses.	
		6. Submit the training management report to the general management when required.	

Conclusions

Proposing an OSHMS is a process that any company, regardless of the industry, must undergo if it wants to control its OSH risks and improve its work performance.

When developing the initial diagnosis it was determined that the level of compliance with the OHSAS standard in the sanitation company SEDA-JULIACA is 30%, which indicates that this company does not have a good management system, so it is necessary to implement an OHSMS according to the required legal provisions.

The processes involved in the recognition of hazards and risk assessment are aimed at integrating and demonstrating compliance with the OSHMS, which guides the employees involved in the various operations performed within the company SEDA-JULIACA.

Likewise, the IPERC was prepared for each activity carried out at the SEDA-JULIACA treatment plant, which is essential to provide the necessary measures that will be considered in the proposal.

Suggestions

To the companies, it is suggested to describe in a simple and easy-to-understand way the OSHMS policy and goals, since it is indispensable for the constant progress of the organization.

It is also necessary to specify accidents and illnesses at work in order to carry out preventive actions, and it is also essential to implement an emergency plan.

It is also recommended that a technical study be made according to the PPE needs of each work station and that workers be trained in their use and protection.

In addition, it is suggested to train personnel on an ongoing basis, as well as to update international and national standards for managers and personnel assigned to the leadership of occupational risk prevention.

CHAPTER FIVE :

**EMERGENCY PLAN AT THE CORPORATE LEVEL:
AN OBLIGATION OR A PREVENTIVE MEASURE?**

Over the years, various natural disasters and pandemics have occurred, which have led to economic and social crises, etc. In this sense, entities responsible for risk management have been developed, with the purpose of reducing the level of criticality and impact on the population, to be prepared for future contingencies and to increase productivity in each country[61] .

Therefore, it is necessary for companies to design a plan "to prevent and anticipate partial or total disasters"[62] , such as fires, earthquakes, floods, among others, and to provide strategies that allow those responsible to have information on how to act before, during and after these contingencies occur.

For this reason, the emergency plan for the company must be a priority, which must take into account the policies or processes to respond immediately to the situation occurring in the company. This is determined not only by current regulations, but also by the requirements of each company and the environmental and social conditions to which it is exposed[63] .

---

[61] World Economic Forum. *The Global Risks Report 2023*, 18th ed., Switzerland, Worl Economic Forum, 2023, available at [https://es.weforum.org/reports/global-risks-report-2023].

[62] Cristian Giovanni Rodríguez Rodríguez. "The importance of a business continuity plan." *Universidad Piloto de Colombia*, vol. 1, pp. 1-10, 2020, available at [http://repository.unipiloto.edu.co/handle/20.500.12277/9547], p. 1.

[63] María Alejandra Orjuela Perdomo and María Alejandra Ruge Vera. "Propuesta de implementación del plan de emergencias y contingencias para la empresa Inversiones Jomayosa SAS basado en la norma 45001:2018" (Master's thesis). Bogotá, Colombia, Universidad ECCI, 2021, available at [https://repositorio.ecci.edu.co/handle/001/1289].

114

According to Bello *et al.*[64] , a plan for these situations should include:

— a preferred method of reporting fires or other emergencies;

— evacuation policy and procedures;

— assignment of escape routes and modes of escape in an emergency (area maps, recognition of emergency or refuge zones);

— data of each personnel of the emergency brigade;

— procedures that employees must follow to perform or stop critical operations in their area, to operate fire extinguishers or to evacuate upon hearing an emergency alarm;

— and rescue or medical actions for designated employees.

Its importance lies in the fact that it optimizes employee preparedness and response capacity by providing first aid, reduces vulnerability to emergency situations from trained employees, and increases technical knowledge through the use of practical learning materials based on recreational practices.

Other essential benefits focus on motivating employees to participate in the various disaster intervention tasks, building a peaceful and safe working environment, minimizing the impact and severity of potential disasters and thus avoiding human and economic losses[65] .

---

[64] Omar Bello, Alejandro Bustamante and Paulina Pizarro. *Planning for disaster risk reduction within the framework of the 2030 Agenda for Sustainable Development*. Santiago, ECLAC, 2021, available at [https://repositorio.cepal.org/server/api/core/bitstreams/ae6fe59f-e288-431b-8edd-7cbe1f760c8d/content].

[65] María Alejandra Orjuela Perdomo and María Alejandra Ruge Vera. "Propuesta de implementación del plan de emergencias y contingencias para la empresa Inversiones Jomayosa SAS basado en la norma 45001:2018", cit.

BIBLIOGRAPHY .

Abril Sánchez, Cristina; Antonio Enríquez Palomino and José Manuel Sánchez Rivero. *Guía para la integración de sistemas de gestión: calidad, medio ambiente y salud en el trabajo*, 2nd ed., Madrid, Spain, Fundación Confemetal, 2012.

Achinte Hurtado, Adriana Stella and Sidney Oriana Henao Clavijo. "Planning of the occupational health and safety management system for a locative maintenance company based on Decree 1072 of 2015, period 2015-2016" (specialty thesis). Cali, Colombia, Universidad Libre, 2016, available at [https://repository.unilibre.edu.co/handle/10901/9893?show=full].

Agustini Paredes, Liliana Rosalinda; Pedro Pablo Rosales López and Anwar Julio Yaroin Achachagua. *Ratios of accidentability*. Lima, Perú, Universidad Nacional Mayor de San Marcos, 2021, available at [https://industrial.unmsm.edu.pe/wp-content/uploads/2021/04/PSEG103-Ratios-de-Accidentabilidad.pdf].

Alcántara Moreno, Gustavo, "La definición de salud de la Organización Mundial de la Salud y la interdisciplinariedad" (The World Health Organization's definition of health and interdisciplinarity). *Sapiens, Revista Universitaria de Investigación*, vol. 9, no. 1, 2008, pp. 93-107, available at [https://www.redalyc.org/pdf/410/41011135004.pdf].

Spanish Association for Standardization and Certification (Ed.). *OHSAS 18001:2007. Occupational health and safety management systems - Requirements*. Madrid, AENOR, 2007, disponible en [https://infomadera.net/uploads/descargas/archivo_49_Sistemas%20de%20gesti%C3%B3n%20de%20seguridad%20y%20salud%20OHSAS%2018001-2007.pdf].

Barrera-García, Aníbal; Alejandro González-Delgado and Damayse Pérez-Fernández. "Identification of incident factors in occupational accident rates in companies in

Cienfuegos". *Industrial Engineering*, vol. 37, no. 2, 2016, pp. 127-137, available at [https://www.redalyc.org/articulo.oa?id=360446197003].

Bello, Omar; Alejandro Bustamante and Paulina Pizarro. *Planning for disaster risk reduction within the framework of the 2030 Agenda for Sustainable Development.* Santiago, ECLAC, 2021, available at [https://repositorio.cepal.org/server/api/core/bitstreams/ae6fe59f-e288-431b-8edd-7cbe1f760c8d/content].

Bernal Mateus, María del Carmen. *La norma OHSAS 18001 y su implementación*, 2nd ed., Bogotá, Colombia, ICONTEC, 2009.

Bustamante Granda, Fernando. "Sistema de gestión en seguridad basado en la norma OHSAS 18001 para la empresa constructora eléctrica IELCO" (Master's thesis). Ecuador, Universidad Politécnica Salesiana, 2013, available at [https://dspace.ups.edu.ec/handle/123456789/5375].

Carrera Endara, Carlos Fernando Atahualpa; Cristian Heriberto Ligña Cumbal, Galo Renan Moreno Cueva and Ruben Morales Carrera. *Quality management systems.* United States, Compás, 2018, available at [http://142.93.18.15:8080/jspui/bitstream/123456789/466/3/SISTEMAS%20DE%20GESTI%C3%93N%20DE%20LA%20CALIDAD.pdf].

Congress of the Republic. *Law of Safety and Health at Work*. Law No. 29783 of 20-08-2011. Lima, Perú, Congreso de la República, 2011, available at [https://web.ins.gob.pe/sites/default/files/Archivos/Ley%2029783%20SEGURIDAD%20SALUD%20EN%20EL%20TRABAJO.pdf].

Congress of the Republic. *Law that amends Law 29783, Law of Safety and Health at Work*. Law No. 30222 of 11-07-2014. Lima, Peru, Congreso de la República, 2014, available at [https://leyes.congreso.gob.pe/Documentos/Leyes/30222.pdf].

CONICYT. *Standards manual: biosafety and associated risks*. Chile, Fondecyt-CONICYT, 2018.

Contri Campanelli, Leandro and Lucas Desiderio Ribeiro. "Involvement of Brazilian companies with occupational health and safety aspects and the new ISO 45001:2018." *Production*, vol. 31, 2021, pp. 1-13, available at [https://doi.org/10.1590/0103-6513.20210005].

Cortés Díaz, José María. *Seguridad e higiene del trabajo: Técnicas de prevención de riesgos laborales*, 12th ed., Madrid, Spain, Tébar, 2012.

Cuba Miranda, Ramiro and César Mercado Rivero. "Implementación de un plan de seguridad y salud ocupacional en las labores de mantenimiento, planchado y pintura en la empresa Fátima Car Service Srl - Cusco - 2021" (undergraduate thesis). Cusco, Peru, Universidad Continental, 2022, available at [https://hdl.handle.net/20.500.12394/11814].

Del Pezo De la Cruz, Otto Gabriel. "Modelo de gestión de seguridad y salud ocupacional para la empresa de agua potable, aguas de Península - Aguapen S. A." (Master's thesis). Ecuador, Universidad Politécnica Salesiana, 2013, available at [https://dspace.ups.edu.ec/handle/123456789/4829].

Strategic Development Management. *Application of the Deming cycle or PDCA for quality management in higher education: an introduction*. Chile, Universidad de Concepción, 2020, available at [https://desarrolloestrategico.udec.cl/wp-content/uploads/2021/01/DDD-N-4-Ciclo-Deming.pdf].

ENEL. *Internal regulations for occupational health and safety*. Lima, Peru, ENEL, 2021, available at [https://www.enel.pe/content/dam/enel-pe/sostenibilidad/sistemas-de-gesti%C3%B3n/enel-distribuci%C3%B3n/sistemas-de-gesti%C3%B3n-actualizados/Reglamento%20Interno%20de%20Seguridad%20y%20Salud%20en%20el%20Trabajo%20-%20V9.pdf].

EPS SEDA-JULIACA. *General company data*. Juliaca, Peru, EPES SEDA-JULIACA, 2023, available at [https://SEDA-JULIACA.com/datos-dela-empresa/].

EPS ILO S.A. *Disposiciones que regulan el régimen disciplinario y procedimiento sancionador de la EPS ILO S.A.* Lima, Perú, EPS ILO S.A., 2020.

Gadea García, Adrián Wilfredo. "Propuesta para la implementación del sistema de gestión de seguridad y salud en el trabajo en la empresa SUMIT S.A.C." (Bachelor's thesis). Lima, Peru, Universidad de Lima, 2016, available at [https://repositorio.ulima.edu.pe/bitstream/handle/20.500.12724/3497/Gadea_Garcia_Adrian.pdf?sequence=1&isAllowed=y].

González González, Nury Amparo. "Diseño del sistema de gestión en seguridad y salud ocupacional, bajo los requisitos de la norma NTC-OHSAS 18001 en el proceso de fabricación de cosméticos para la empresa Wilcos S.A." (bachelor's thesis). Bogotá, Colombia, Pontificia Universidad Javeriana, 2009, available at [http://hdl.handle.net/10554/7232].

Henao Robledo, Fernando. *Salud Ocupacional: conceptos básicos*, 2nd ed., Bogotá, Colombia, Ecoe Ediciones, 2010.

Hernández Domínguez, Juan Daniel. "Occupational safety to prevent accidents in industry." *Ibero-American Journal of Academic Production and Educational Management*, vol. 5, no. 10, 2018, pp. 1-9, available at [https://www.pag.org.mx/index.php/PAG/article/view/773/1109].

Hernández-Sampieri, Roberto and Christian Paulina Mendoza Torres. *Research methodology: the quantitative, qualitative and mixed routes*. Mexico City, McGraw-Hill Interamericana Editores, 2018.

INACAL. *Safety Signs. Graphic symbols and safety colors. Part 1: Rules for the design of safety signs and safety stripes*. NTP 399.010-1 of 29-12-2016. Lima, Peru, INACAL, 2016, available at [https://minercode.org/normastecnicasperuanas/399010-1-2016.pdf].

National Institute of Statistics and Informatics. *Estado de la población peruana 2020.* Lima, Peru, INEI, 2020, available at [https://www.inei.gob.pe/media/MenuRecursivo/publicaciones_digitales/Est/Lib1743/Libro.pdf].

National Institute of Statistics and Informatics. *Síntesis Estadística 2016*. Lima, Peru, INEI, 2016, available at [https://www.inei.gob.pe/media/MenuRecursivo/publicaciones_digitales/Est/Lib 1391/libro.pdf].

Lazo González, Silvia Estefanía. "Implementación de sistema de gestión de seguridad y salud en el trabajo para la realización de lasaña tradicional en la empresa Pizzerias Presto en base a Ley N.° 28783 Ley de Seguridad y Salud en el Trabajo" (Bachelor's thesis). Arequipa, Peru, Universidad Católica de Santa María, 2016, available at [https://repositorio.ucsm.edu.pe/handle/20.500.12920/5518].

Mancera Fernández, Mario; María Teresa Mancera Ruíz, Mario Ramón Mancera Ruíz and Juan Ricardo Mancera Ruíz. *Safety and Industrial Hygiene. Gestión de riesgos*. Colombia, Editorial Alfaomega, 2012, available at [https://ashconsultores.com.ar/wp-content/uploads/2019/06/Libro_Seguridad_e_Higiene_industrial_ges.pdf].

Matabanchoy Tulcán, Sonia Maritza. "Salud en el trabajo." *Universidad y Salud*, vol. 14, no. 1, 2012, pp. 87-102, available at [https://revistas.udenar.edu.co/index.php/usalud/article/view/1270].

Mejía, Christian; Matlin Cárdenas and Raúl Gomero-Cuadra. "Notification of occupational accidents and diseases to the Ministry of Labor. Peru 2010-2014." *Revista Peruana de Medicina Experimental y Salud Pública*, vol. 32, no. 3, 2015, pp. 526-531, available at [https://doi.org/10.17843/rpmesp.2015.323.1689].

Ministry of Energy and Mines. *Decreto Supremo que aprueba el Reglamento de Seguridad y Salud Ocupacional y otras medidas complementarias en minería*. Decreto Supremo N.° 055-2010-EM de 01-01-2011, Lima, Peru, MINEM, 2011, available at [https://www.minem.gob.pe/minem/archivos/file/Mineria/LEGISLACION/2010 /AGOSTO/DS%20055-2010--EM.pdf].

Ministry of Energy and Mines. *National Electricity Code: use.* Lima, Peru, MINEM, 2006, available at [http://www.pqsperu.com/Descargas/NORMAS%20LEGALES/CNE.PDF].

Ministry of Labor, Employment and Social Security. *Occupational Health and Safety (OSH). Aportes para una cultura de la prevención.* Argentina, ILO, 2014, available at [https://www.ilo.org/wcmsp5/groups/public/@americas/@@ro-lima/@ilo-buenos_aires/documents/publication/wcms_248685.pdf].

Ministry of Labor and Employment Promotion. *Approval of occupational health and safety regulations.* Decreto Supremo N.° 009-2005-TR de 29-09-2005, Lima, Peru, MTPE, 2005, available at [https://oiss.org/wp-content/uploads/2018/11/12-03_Reglamento_de_Seguridad_y_salud_en_el_trabajo_2005-09-29_009-2005-TR_487.pdf].

Ministry of Labor and Employment Promotion. *Ley de Seguridad y Salud en el Trabajo, its regulations and amendments.* Lima, Peru, MTPE, 2017, available at [https://cdn.www.gob.pe/uploads/document/file/349382/LEY_DE_SEGURIDAD_Y_SALUD_EN_EL_TRABAJO.pdf].

Ministry of Labor and Employment Promotion. *Notifications of occupational accidents, hazardous incidents and occupational diseases.* Lima, Peru, MTPE, 2022, available at [https://cdn.www.gob.pe/uploads/document/file/4327880/SAT_DICIEMBRE_2022.pdf?v=1679929130].

Ministry of Labor and Employment Promotion. *Approval of reference formats that contemplate the minimum information to be contained in the mandatory records of the Occupational Health and Safety Management System.* Ministerial Resolution No. 050-2013-TR of 14-03-2013. Lima, Peru, MTPE, 2013, available at [https://www.gob.pe/institucion/mtpe/normas-legales/288031-050-2013-tr].

Ministry of Housing, Construction and Sanitation. *Decreto Supremo que modifica el Reglamento de Organización y Funciones del Ministerio de Vivienda, Construcción y Saneamiento.* Supreme Decree No. 010-2014-VIVIENDA of 03-

03-2015, Lima, Peru, MVCS, 2015, available at [https://cdn.www.gob.pe/uploads/document/file/360774/DS-006-2015-VIVIENDA.pdf].

Morell González, Luisa María; Rosa Maricela Cedeño Zambrano and Sarai Ramírez Cruz. "Systemic management of institutional risks. Diagnosis in the Agrarian University of Havana and the Technical University of Manabí". *Cofin Habana*, vol. 13, no. 1, 2019, pp. 1-10, available at [http://scielo.sld.cu/scielo.php?script=sci_arttext&pid=S2073-60612019000300016].

Niciejewska, Marta and Olga Kiriliuk. "Occupational health and safety management in 'small size' enterprises, with particular emphasis on hazards identification." *Production Engineering Archives*, vol. 26, no. 4, 2020, pp. 195-201, available at [https://sciendo.com/article/10.30657/pea.2020.26.34].

International Labor Organization. *Occupational safety and health in Peru. Una mirada desde los convenios internacionales del trabajo no ratificado.* Peru, ILO, 2022, available at [https://www.ilo.org/wcmsp5/groups/public/---americas/---ro-lima/documents/publication/wcms_884854.pdf].

International Labor Organization. *Occupational safety and health inspection. Training module for inspectors.* Argentina, ILO, 2017, available at [https://www.ilo.org/wcmsp5/groups/public/---americas/---ro-lima/---ilo-buenos_aires/documents/publication/wcms_592318.pdf].

Orjuela Perdomo, María Alejandra and María Alejandra Ruge Vera. "Propuesta de implementación del plan de emergencias y contingencias para la empresa Inversiones Jomayosa SAS basado en la norma 45001:2018" (Master's thesis). Bogotá, Colombia, Universidad ECCI, 2021, available at [https://repositorio.ecci.edu.co/handle/001/1289].

Ortega Alarcón, Jaime Antonio; Jorge Rafael Rodríguez López and Hugo Hernández Palma. "Importance of worker safety in the compliance of processes, procedures and functions." *Academia & Derecho Journal*, vol. 8, no. 14, 2017, pp. 155-176, available at [https://dialnet.unirioja.es/servlet/articulo?codigo=6713605].

Pastor Fernández, Andrés; Manuel Otero Mateo, José María Portela Núñez and José Luis Viguera Cebrián. *Manual de prácticas de seguridad en el trabajo*. Spain, Editorial UCA, 2016.

Pérez Cabrera, Arturo Miguel. "Incidence of accident risks on the operating costs of non-metallic resource mining concessions in Patapo - Lambayeque" (undergraduate thesis). Pimentel, Peru, Universidad Señor de Sipán, 2019, available at [https://repositorio.uss.edu.pe/handle/20.500.12802/5551].

Pérez, José Luis. "Sistema de gestión en seguridad y salud ocupacional aplicado a empresas contratistas en el sector económico minero metalúrgico" (Master's thesis). Lima, Perú, Universidad Nacional de Ingeniería, 2007, available at [http://hdl.handle.net/20.500.14076/633].

Presidency of the Republic of Peru. *Approve the Law on Safety and Health at Work*. Supreme Decree No. 005-2012-TR of 01-11-2016. Lima, Peru, Presidency of the Republic of Peru, 2016, available at [https://www.gob.pe/institucion/presidencia/normas-legales/462577-005-2012-tr].

Purga Ruiz, Wendy Arelí and Anthony Percy Torres Vargas. "Propuesta de implementación de un sistema de seguridad y salud ocupacional basado en la norma OHSAS 18001:2007 para evitar costos por incidentes en el consorcio Alvac Johesa" (undergraduate thesis). Lima, Peru, Universidad Privada del Norte, 2017, available at [https://repositorio.upn.edu.pe/handle/11537/12390].

Raffo Lecca, Eduardo. *Introducción a la seguridad y salud en el trabajo*. Lima, Colecciones Jovic, 2016.

Rodríguez Mesa, Rafael. *Sistema general de riesgos laborales*, 3rd ed., Colombia, Universidad del Norte, 2017.

Rodríguez Rodríguez, Cristian Giovanni. "The importance of a business continuity plan." *Universidad Piloto de Colombia*, vol. 1, pp. 1-10, 2020, available at [http://repository.unipiloto.edu.co/handle/20.500.12277/9547].

Romero Albán, Ángela Liliana. "Diagnóstico de normas de seguridad y salud en el trabajo e implementación del reglamento de seguridad y salud en el trabajo en la Empresa Mirrorteck Industries S.A." (Master's thesis). Ecuador, Universidad de Guayaquil, 2014, available at [http://repositorio.ug.edu.ec/handle/redug/4494].

Sabogal Barbosa, Maritza Edilma and Félix Eduardo Rodríguez Medina. "Competencies of the technician in the work area. Un análisis del programa técnico manejo de prevención de riesgos laborales de la Fundación Universitaria San Mateo." *Plataforma Abierta de Libros y Memorias Académicas*, vol. 1, 2019, pp. 9-36, available at [https://cipres.sanmateo.edu.co/ojs/index.php/libros/article/view/396].

SUNAFIL. *Report for the transfer of management of the National Superintendence of Labor Inspection - SUNAFIL. Period of government 2011 - 2016*. Lima, Peru, SUNAFIL, 2016.

Terán Pareja, Itala Sabrina. "Propuesta de implementación de un sistema de gestión de seguridad y salud ocupacional bajo la norma OHSAS 18001 en una Empresa de Capacitación Técnica para la Industria" (Bachelor's thesis). Lima, Pontificia Universidad Católica del Perú, 2012, available at [http://hdl.handle.net/20.500.12404/1620].

Tirado Medina, Jefferson Andrée and Víctor Luis Vega Ybáñez. "Propuesta para la implementación de un plan de seguridad y salud ocupacional para controlar los riesgos y reducir los accidentes en la división de mantenimiento de la empresa de servicio de agua potable y alcantarillado de la Libertad - Sedalib S.A." (Bachelor's thesis). Trujillo, Peru, Universidad Nacional de Trujillo, 2017, available at [http://dspace.unitru.edu.pe/handle/UNITRU/8880].

Valladarez Tola, Jaime Mauricio. "Implementación del sistema de gestión en seguridad y salud ocupacional bajo una nueva versión de la norma OHSAS 18001:2007 en la Corporación Eléctrica de Ecuador Celec-Hidropaute" (Master's thesis). Ecuador, Universidad de Cuenca, 2010, available at [http://dspace.ucuenca.edu.ec/handle/123456789/2631].

Vásquez Ojeda, Marco Antonio. "Implantación de un sistema de gestión de seguridad y salud ocupacional en el proyecto especial Olmos - Tinajones - Lambayeque" (Master's thesis). Trujillo, Peru, Universidad de Trujillo, 2016, available [http://dspace.unitru.edu.pe/items/62188267-261a-49a1-bf9d-9ad96b750193].

World Economic Forum. *The Global Risks Report 2023*, 18th ed., Switzerland, Worl Economic Forum, 2023, available at [https://es.weforum.org/reports/global-risks-report-2023].

Printed by Books on Demand GmbH, Norderstedt / Germany